SMELLING TO SURVIVE

SMELLING TO SURVIVE

The Amazing World of Our Sense of Smell

Bill S. Hansson

HERO, AN IMPRINT OF LEGEND TIMES GROUP LTD
51 Gower Street
London WC1E 6HJ
United Kingdom
www.hero-press.com

Originally published in German as *Die Nase vorn. Eine Reise in die Welt des Geruchssinns* by S. Fischer Verlag in 2021
First published in English by Hero in 2022

 British Library Cataloguing in Publication Data available.

Front cover design and illustration by Laura Brett

Printed and bound by CPI Group (UK) ltd, Croydon, CR0 4YY

ISBN: 978-1-91505-449-4

Contents

Introduction

It's spring and the fields are freshly ploughed. There's a very special, pleasant smell in the air. If you've ever experienced such a moment, the smell signifies exactly this situation in your brain. Springtime, open soil, farmland. Maybe you're thrown back in time by a memory you didn't even know you had. Few sensory experiences are as good at recalling earlier experiences as olfactory ones (smell). It seems like the memories are simply lying in wait for the right odour to trigger them once again.

One of the most powerful examples in literature of the memory-unlocking ability of smell is found in the first of Marcel Proust's seven-volume masterpiece *In Search of Lost Time*. It opens with the sweet smell of madeleines, miniature sponge cakes, evoking involuntary memories of the author's childhood and adult life. But smell is not a sense that is unique to humans.

All organisms, with and without backbones, from insects to humans, use sensory systems to make sense of their environment and to communicate with each other. During the course of evolution, different species have become more or less dependent on a certain kind of information. Crickets and bats rely heavily on sound waves, dragonflies and humans often put their trust in sight, while moths, pigs and dogs are famous for their keen sense of smell.

As humans are indeed very visual beings, we tend to forget the other senses. Especially our own sense of smell. This is partly because we are less dependent on chemical information nowadays. But there is also something quite primitive about smell. Something we want to avoid. Just think about the lengths we go to in order to disguise our own, natural smells, to cover them with artificial odorants or to prevent them with deodorants. We may think we are less dependent on olfactory information than other species, but we aren't really. Many vital aspects of our lives depend heavily on smells. I will explore why and how in my chapter about the human sense of smell.

For other animals, an acute sense of smell is absolutely vital for survival and reproduction. Back in the 1800s, when the French entomologist Jean-Henri Fabre noticed how huge numbers of male moths were attracted to a caged female in his house, he assumed that odours were involved. We now know that he was on the right track. The male moth can follow the female scent, a trail close to homeopathic concentration given off by the female, making him probably the best smeller of them all.

When salmon returns to spawn in the same river branch where it was born, it uses smell to find its way. Without its sense of smell, it would be lost. So specific are the odours in the water that each river tributary has its own signature. Male dogs are as keen as moths, if not as sensitive, to find the bouquet of a female in heat. Even so, dogs are a thousand times more sensitive to smell than we humans. It's an ability that we've put to good use in many contexts, including for hunting and tracking, for locating earthquake victims and even for diagnosing cancer. For dogs, life is lived much more in a landscape of odours rather than visuals. As such, dogs "see" history as odours, not as visual impressions. Smells linger on and can tell them what happened – or who passed by – long after anyone saw it.

For a long time, birds were believed to have no or a very poor sense of smell. Today, we know otherwise. Vultures can pick up the scent of the distinct molecules emitted from a dead animal from far, far away. While seabirds, such as albatrosses, can smell their way to an ample supply of plankton, which in turn means good opportunities for fishing.

What's even more surprising perhaps is the fact that plants can smell and send odour messages to each other. They also use specific odours to manipulate friends and foes. When a plant is attacked, for instance by moth larvae, it changes its emission of volatiles. These molecules can have two different positive effects for the plant. They warn neighbours of the same species that an attack is going on, so that they can turn on their defence system before the herbivores get to them. The volatiles can also serve as a "call for help" by attracting enemies of the attackers. Your enemy's enemy is your friend, even in the plant world.

In another context, plants have evolved to attract the insects they are dependent on for pollination. Normally, this is a process that is a win-win for both players. Sometimes, however, the plant cheats the insects into doing the job without any payback whatsoever.

From all these examples, it's obvious that most organisms on earth are dependent on odour information to survive and reproduce. Being able to sense your chemical environment allows you to adapt to the surrounding conditions, find a resource or a mate and avoid different types of enemies, toxic substances and disease agents.

Before we can understand how smell works, we need to understand what smell really is. Both smell and taste consist of chemical information. The molecules in a water solution give us taste, while in air they give us smell. For something to smell it has to emit molecules light enough to take flight. A piece of sugar doesn't smell, as the molecules are too heavy to take off. While no one would mistake the molecules escaping from a lemon for anything else. The limonene and citral molecules fly easily towards our nose.

All emitted molecules however are not odours. Only when they can be detected by another organism do they contribute to the smell, for example, of a banana. The number of molecules emitted is impressive. A banana sends out hundreds of different molecules. Only a few of these are indeed odours detected by an insect or the human nose, while all the others are just volatile molecules.

To detect smells, all animals need some kind of detector system. To perform this task a specific part of the nervous system has to make contact with the environment and must be endowed with specific receptors to recognize relevant molecules. Our nose is actually the only place where our nervous system is in direct contact with our surroundings. The nerves hang out into the environment. Well, not quite, as they swim in a sea of snot inside your nose. But still, they are exposed to all kinds of pollutants and dust that enter the nose together with smells. Nerves can, however, neither see nor smell. They need to be equipped with some detectors, so-called receptors, to perform.

To see, humans need only three receptor types to register all visible light. All light consists of a waveform oscillating faster or slower, which gives the impression of different colours.

When it comes to smelling, the situation is very different. Every type of odour molecule has a unique chemical property very different from all other molecules. That's why we don't have only three but around 400 olfactory receptors. Otherwise we wouldn't be able to smell the millions of different odours that we can distinguish between. Most receptors can detect a spectrum of different molecules. Their activation is similar to playing a piano. With 400 receptor keys to press, millions of odour melodies can be played.

Once the molecules of smell have been detected by the nerves in your nose, signals travel to a specific area of your brain where information is organized into little balls, glomeruli, of nervous tissue. Each glomerulus receives input from nerves carrying a specific receptor type. This means that the "melody" is translated into a three-dimensional map of activity. This map is read out by the next levels of neurons and in the end transferred to other areas in the brain, such as the hippocampus and the amygdala, where the significance of the odour is encoded and put into context. I will come back to the importance of these areas and the complete system.

Interestingly, the basic architecture of the olfactory system is very similar in most organisms studied (excluding plants). The peripheral nerves with receptors converge on little balls of nervous tissue and finally target specific brain areas. In animals as distinct as flies and humans we see the very same building blocks.

In more or less all animals, the sense of smell thus has a similar architecture, even though it no doubt has different evolutionary origins. Convergent evolution has probably made it quite similar all the way from insects to humans. To smell, the nose of all organisms needs to be equipped with some kind of chemical detectors: neurons that are able to detect different kinds of molecules in the air (or for fish in the water). This detection and identification of molecules happens in olfactory receptors residing in the membrane of the smell nerves, the olfactory sensory neurons.

The receptors consist of proteins traversing the membrane of the neuron seven times and thereby forming pockets and folds, where the smell molecules can fit like keys in a lock. When a key fits, it unlocks a cascade of neurochemical events called a transduction cascade, which in the end causes the neuron to react electrically. This signal can then travel via the neuron's axon to the first olfactory station of the brain.

But before we move into the brain, let's look at the microenvironment around the olfactory sensory neurons. In the noses of all mammals, birds and other land-living vertebrates, the neurons hang out straight into the air. This is the only place on our body where neurons are actually directly exposed to the environment. The nose has therefore been equipped with a protective mucus layer surrounding the exposed neurons. In insects and other arthropods, the neurons have been encased into small hairs on the antennae and the palps (the insect noses). Each little hair also contains mucus. This mucus has about the same composition as seawater but with a lot of proteins added, making it thick and less prone to evaporate. These proteins also help in dissolving fatty molecules in the nose's ocean water.

From the antennae and the nose, the olfactory sensory neurons project their axons to the olfactory bulb (vertebrates) or the antennal lobe (arthropods) of the brain. In all the animals described here, these primary olfactory brain centres have a more or less similar architecture. The axons of the nose neurons find their way to the glomeruli. Each olfactory sensory neuron type, expressing a specific type of olfactory receptor, targets one little glomerulus of the bulb/lobe. This means that when neurons in the nose or the antennae are activated, a map of activity will be painted over the glomeruli. In insects, we typically find 50–500 glomeruli, while a mouse, for example, has around 2,000 and a human even more.

Within the olfactory bulb or antennal lobe, some information processing is going on thanks to widespread local neurons shuffling information from one little ball to the other, allowing the input from different types of odours to affect each other. In the end, the processed message leaves the lobe/bulb via neurons targeting higher brain areas involved in perception, memory, decision-making or other cognitive processes.

What about all the odour messages flowing between and within species? Well, there is special terminology for all these semiochemicals. You will find these repeated in many of the chapters, but let's take a quick look at them here first.

An odour that is sending a message between individuals of the same species is called a **pheromone**. A typical example is when a female dog in heat sends out an odour message that calls on every male dog in the vicinity and the message is: "Come mate with me!" You will see many examples of pheromones in the coming chapters.

The remaining semiochemicals send messages between species. They are typically divided into categories depending on who benefits from them, the sender or the receiver. If the receiver benefits, they are called **kairomones**. A typical example would be the odour given off by a prey animal, for example a mouse, and picked up by a predator, often a cat.

If the smells benefit the sender, they are called **allomones**. Any kind of lure would fall into this category, but it would also include a defence mechanism, such as a skunk sending out a stinky spray to fend off an enemy.

Finally, an odour message can also benefit both parties. In this case, it is called a **synomone**. The classic example here is the smell of insect-pollinated flowers, where both the flower gets pollinated and the insect receives a reward in the form of nectar and pollen.

All the information we as humans have gathered on how the sense of smell works, which molecules are involved and which types of behaviours are displayed in response to these, allows us to design different strategies to assist us in different ways. Today, electronic noses play an important role in disease diagnosis, security checks and surveillance of environmental pollution. No one can avoid the huge industry involved in inventing new and alluring odours for us to apply over our bodies. When the pig breeder wants to inseminate the sow, he buys synthetic boar pheromone to get her in the mood. Insects of many different kinds are managed using pheromones and plant odours.

In this book, I will use different examples from the world around us to describe the fascinating world of smells. An understanding of our own olfactory system, its function and its architecture, provides an important foundation before we venture out into all other systems. In a number of chapters I will tell captivating stories emerging from my own and my colleagues' research. The stories will cover different animals but will also look at how plant smell affects our environment. I will begin by exploring how climate change might affect the ecology of smell – and will end with an overview of how humans make use of all our knowledge on smell and odour-guided behaviours to our own benefit.

SMELLING TO SURVIVE

Chapter 1

Smelling in the Anthropocene

If you were walking down the road 1,000 years ago your sensory experience would probably be quite different to the one you have today. Looking around you in 1022, you would see no cars, airplanes or ships. Maybe not even a proper road in the modern sense of the word.

Undoubtedly, the world would be a whole lot quieter, almost silent perhaps. These are our impressions from sound and vision, but what about from our sense of smell?

There are so many levels to the sense of smell. Do we and our environment smell different(ly) today compared to a millennium ago? Or even 100 years ago? How exactly have the scents in our surroundings changed over the years? How have we humans contributed to this changing smellscape – the complex landscape of odours and aromas around us? Have our own smells and perception of odours changed along the way? How have our activities affected our capacity to smell? What actions are to blame for bringing about such changes in both humans and animals?

Well, for a start, in 1022 you couldn't expect to get a whiff of a car exhaust or a stink from the local water treatment plant. You would not be exposed to synthetic odours either: perfumes, deodorants or that new car smell, for example. Even natural odours might have been different.

Ever since humans started colonizing Earth's every corner, we have found ways to change, manipulate and exploit our environment. To name just a few: we have cut down forests, planted crops, exterminated both plants and animals and industrialized the world. This new geological epoch, where the world has been changed dramatically through human activities, is often referred to as the Anthropocene.[1]

A clear definition of the exact time span of this period is still disputed. Suggestions for its starting point range from the onset of the agricultural revolution, around 10,000–15,000 years ago, to just after World War II, a period defined by nuclear tests, the post-1950 Great Acceleration and the accompanying dramatic socioeconomic and climate changes.

Whichever scale we choose, it's clear that humans have had an immense impact on this planet in general, but also on every breath we and other animals take. As well as on the very molecules contained in each of those sniffs.

Our changing smellscape

Let's consider natural smells first, and how they might have changed. A thousand years ago, nature was still quite unaffected by humans. Many species of plants and animals coinhabited fields and forests. Flowers were abundant. Pine and spruce were mixed with many species of deciduous trees. The keyword was diversity. As time went by, humans cut and burnt down forests and transformed flowering meadows into farm fields. All these changes allowed the great spread and multiplication of the human race. At the same time, they altered the smellscape around us profoundly.

Instead of diverse, mixed-species forests, we got large-scale, single-species tree cultivations. In the same way, smells were simplified. Take for instance the scent of a modern spruce forest compared to an old mixed stand. If you get the chance, try making the comparison yourself the next time you're out there in the woods.

In parallel, the same simplification has been underway in the fields. What were once great mixes of species have become huge monocultures. The US prairie turned into never-ending corn and wheat fields. European meadows have gone the same way. When contemplating the so-called natural odours around us, the smellscape has, already, gone through a pronounced change. How so?

The disruptive role of CO2

When we drive or fly or operate our industries, we emit many substances that also tend to affect the climate and the molecules carried in the atmosphere. One of the most publicized changes associated with the Anthropocene is an increased level of environmental CO_2, which contributes to the greenhouse effect, the dramatic shift in world temperatures, increased acidity of our oceans and the overall destabilization of our climate.[2]

Although CO_2 is a rather non-reactive compound with no direct chemical impact on odours in the atmosphere, ambient CO_2 can modify plant emission of volatile compounds. This happens through physiological changes within the plant. Carbon dioxide can increase a plant's photosynthetic activity by reducing water consumption and by changing the chemical composition of the plant tissues.[3] Variations in CO_2 levels can also affect the ability of insects to locate their hosts. Moths track CO_2 bursts at flower openings to locate their nectar sources. Impaired flower targeting in elevated CO_2 therefore impacts both pollination and pest infestation.[4]

Elevated backgrounds of CO_2 reduce the ability of a mosquito to locate a blood host, as CO_2 is one of the major olfactory cues used by mosquitoes in host detection (see Chapter 9).[5] This might be considered a benefit from a human perspective, but there are downsides.

From an evolutionary perspective, the rate of mosquito speciation has been shown to increase dramatically during periods of elevated levels of atmospheric CO_2.[6] This increased speciation rate may have been driven by the reduction in the CO_2 host signal quality, which has led to other, more specific, smells to function as potential isolating mechanisms between new species. As such, the projected rise in atmospheric CO_2 from anthropogenic activity has important implications for human health, and potentially pollination efficiency, arising from changing insect abundance and distribution.

On land, the outlook is grim. At sea, it's just as bad. CO_2 dissolves in the oceans, forming carbonic acid, which makes the water more acidic.[7] Studies have also shown that acidic water disturbs the sense of smell in marine organisms. Whether they use this sense to detect and avoid predators, to locate food or to track down a mate, a lower ocean pH is likely to disrupt marine life considerably and make these tasks more difficult.[8] It's not yet known whether the marine ecosystem and the food web can adapt to these changing conditions.

Gases galore and shifts in temperatures

Unlike CO_2, ozone and nitrogen oxide (NOx) can directly affect odour blend composition due to their oxidative power. Recently, both pollutants have increased in the atmosphere, and are expected to increase even further.[9] With increasing amounts of these pollutants, the odour blends insects use to locate food, hosts or oviposition (egg-laying) sites are ever more likely to change. While each of these aspects has individual effects, interactions among them will result in other effects, too.

NOx gases are produced whenever we burn different kinds of fuels. They are health hazards per se, but they also cause acid rain and smog. Nitrous oxide, known as laughing gas, also adds to global warming. Methane is produced in many natural processes, including the frequently cited cow farts and burps. But now it is also being released from the thawing tundra, ecologically the coldest of all the biomes, further adding to record high temperatures.

Ozone in the upper atmosphere forms a natural protective layer around the Earth, shielding us from sun radiation. At ground level, however, it is the main constituent of smog. It's formed when sunlight interacts with different types of emissions from human activities.

On top of all these different gases, we pile on many types of herbicides, fungicides and insecticides to control problematic weeds, fungi and insects. These chemicals have been shown to affect olfaction. And finally, human activities tend to release metal ions that can have a direct impact on the olfactory senses.

Shifting air and sea temperatures are key features of the Anthropocene. Will they influence the way we smell the world? While increased ambient temperature could directly affect odour composition, as the amount of each compound in a blend is a function of its volatility, it could also indirectly upset the physiological response of both the emitter and the receiver.

CHAPTER 1

The insect world

In recent years, considerable attention has been raised by studies revealing that we are losing our insects. In some areas of Germany, for instance, insect biomass has decreased by more than half.[10] Such a dramatic change in our biotic environment has some pretty severe consequences for humans. Bee populations are dwindling, which means that fruit trees are not pollinated, and no honey is produced. Bumblebees are also negatively affected, as are several other beneficial insects.

What's more, as insects form the staple food for many of our birds, these creatures are also suffering from a food shortage. Could this decrease in insect numbers be caused by gas and pollutant effects on olfaction and odours? This seems, at least partly, to be a possibility. In several studies of different systems, it has been shown that the gases we emit cause smells to change.

Let's take insect pollination as an example. Over millions of years, co-evolution has fine-tuned the interaction between flower and insect for the benefit of both (well, most of the time – see Chapter 13). Flowers have a visual appearance that insects use for more long-distance orientation, while floral aromas guide the insects on their final approach. When all of this works out, the plant gets pollinated and the insect gets rewarded in the form of nectar and pollen. Still, we're talking about a vulnerable system. We were able to show just how vulnerable by removing the most intimate smell interaction between flower and insect (see Chapter 7 on moths for more details of this research).

If floral scent disappears, no pollination takes place and no nectar is removed. Because of the delicate nature of this system, however, the smell doesn't need to disappear entirely to disrupt communication. It might only need to change. And this is what we see happening after pollution with gases, particularly with ozone.

The ozone effect

Ozone has a very strong oxidizing effect. This means that it causes chemical reactions in other molecules. In my lab we carried out an experiment where we let tobacco sphingid moths fly towards a specific flower in the wind tunnel. First, we replicated current conditions found in nature, and the moths easily located the flower and both pollination and nectar feeding took place. Then, we placed the flower under elevated ozone levels and observed the moth behaviour again. The insect was clearly disoriented and failed to locate the flower.

When we analysed the molecules emitted by the flower it turned out that several of them had changed to something else with a very different smell.

Exposure to an ozone level that already exists during warm days in some parts of the world had a disruptive effect directly on the pollination services provided by the insect. We continued our experiments to see if some plasticity in the insect system could ameliorate the effects of the ozone, and this was indeed what we found.

If a moth was provided with the "new" floral scent along with strong visual guidance, a single experience of the new smell together with a nectar reward was enough for the moth to learn to fly towards the ozonated smell and use it for future feeding.[11] As Ian Malcolm says in *Jurassic Park*: "Life finds a way."

Most examples, however, reveal detrimental effects of high ozone levels on the pollination services of bees, bumblebees, moths and others. The same also holds true for other gases, such as diesel exhaust.[12] It is therefore clear that we should do our utmost to limit the emissions of these gases and preferably decrease them substantially.

In another study, my colleague Geraldine Wright investigated the effects of "modern" pesticides on bee pollinators. Neonicotinoids are the most used insecticides in the world and are less harmful to birds and mammals than the old carbamates and organophosphates. Lower levels were also supposed to be less harmful to the beneficial bees. However, when Geraldine looked at olfactory learning in honeybees that had been exposed only to very low concentrations of neonicotinoids it was clear that they were severely affected.[13] Again, the olfactory communication and underlying abilities were impeded by human interventions.

The role of temperature fluctuations

Temperature also affects the life of insects. A higher temperature will make all smell molecules evaporate a lot faster and everything might smell a bit more. As insects have no thermo-regulation – they lack the ability to maintain a stable body temperature – their physiological functions are often finely tuned to the ambient temperatures of their habitat. Their sense of smell is no exception. A desert-living beetle might have an optimal function at 40°C, while my own recordings from smell neurons in the antennae of winter moths show that these have a temperature optimum of around 10°C. When you reach 20°C the system hardly functions any more. This means that an ever-increasing temperature caused by climate change will have a direct effect on the sense of smell of insects and probably on many other "cold-blooded" animals, too.

A rise in temperature also allows insects to invade new areas of the world. Even though the spread of the insects is not directly connected to olfaction, it is clear that several notorious odour-guided insects are experiencing a boom. In Chapter 9 you will read about the malaria mosquito. This is only one of many species spreading diseases over the world. Right now, we see them invading new areas, including Europe and North America. More recently, the mosquito-borne Zika virus has spread into southern USA from South and Central America, thanks to the spread of *Aedes* mosquitoes. Other diseases, such as West Nile virus and chikungunya, are also on the spread as new areas open up for the vector mosquitoes.[14]

In Chapter 10 you will learn about the smell life of bark beetles. Only a decade ago these beetles produced one generation of offspring, which meant that each female could produce 60 beetles in a year. Now, we have up to three generations in central Europe, which means that a single female can see 3,000 beetle offspring going into hibernation after killing a large number of spruce trees.

Insect research ahead

Clearly, more research is needed to investigate what is going on. In an attempt to understand exactly how the Anthropocene is influencing the smell life of insects, I initiated the Max Planck Center next Generation Insect Chemical Ecology (nGICE for short) to focus specifically on this area. It involves linking and teaming up experts in this broad field across three different institutions: my own Department of Evolutionary Neuroethology at the Max Planck Institute for Chemical Ecology in Germany, the Swedish University of Agricultural Sciences and the Pheromone Group at the Department of Biology at Lund University, also in Sweden.

Our common goal is to uncover how climate change, greenhouse gases and air pollution influence and impact the chemical communication between insects. We want to understand how insects adapt to these changes in their environment. Our aim is to contribute to solving global problems in the context of climate crisis, global nutrition and combating diseases.[15]

Smelling plastic

In 1907, in New York, Leo Baekeland, a Belgian chemist, invented Bakelite, the first plastic made from synthetic components. Since then, the production of plastic has taken on enormous proportions. We have now reached a world output of an estimated 360 million tonnes per year. Why does this matter to olfaction?

As we will see in Chapter 4, birds use their sense of smell for several reasons. For pelagic birds, one important feature of their noses is the ability to smell dimethyl sulfide (DMS). This is a compound emitted by crushed phytoplankton, often when consumed by zooplankton. For birds, the presence of this sulphur gas is therefore a telltale sign that plenty of food is around.

Unfortunately, relying on this molecule to find food also creates a problem in this plastic age. When plastic has been floating in the water for a couple of months it starts to release DMS – deceiving nature into believing it is edible in the process.[16] The UN Environment Programme states that we pour around eight million tons of plastic into the oceans of the world each year[17] – quite possibly adding up to a total of over five trillion macro and micro pieces of plastic (and counting...), there's plenty around to confuse sea life. Birds mistakenly eat the plastic, which clogs their digestive systems and ultimately kills them. This is the reason why an estimated one million seabirds die each year – because their stomachs are full of our plastic debris.

Not only birds have developed the capability to use DMS to find food in the ocean. Both seals and whales (see Chapter 5) are very likely to use the same strategy, exposing themselves to the very same plastic dangers. In one study of baby turtles, 100 per cent of these tiny creatures had plastic in their stomachs.[18] Our immense production of single-use plastics has created such severe environmental repercussions.

In the Great Pacific Garbage Patch (one of five "garbage patches" identified in our oceans), currents and winds accumulate our discarded debris (including plastic and discarded fishing gear) into an area roughly twice the size of Texas – or three times the size of France if you prefer a European scale.[19] Its surface is largely covered by microplastics. These microplastics, studies suggest, may already outnumber zooplankton – and have definitely already made their way down to the Mariana Trench, the deepest spot in our oceans.[20] You can imagine what this trend is doing to birds and other sea-living creatures attracted by the smell.

A sea change in smell

Of course, on top of the airborne smell of DMS that affects birds and animals, there is also man-made chemical pollution spreading through our waterways, oceans, lakes and rivers. Fish, crustaceans and other aquatic creatures are living in a soup of man-made molecules, where some are devastating for them and their ecological system.

Just as in our own system for smelling, fish olfactory sensory neurons are directly exposed, in their case to the surrounding water and everything dissolved in it. Take copper. Studies have shown that increased copper concentration has a direct detrimental effect on the function of fish smell neurons, and likewise on shore crabs and crayfish. Chronic exposure to elevated levels of copper disrupts normal, odour-directed behaviour involved in mating and foraging.[21]

When we protect our crops, we spray pesticides of different kinds that sooner or later find their way to water streams. Most of us with gardens will have at some time or other used glyphosate-based herbicides against weeds. This compound, when tested in concentrations occurring in nature, prevented fish from finding food and directly affected the function of the nose in Coho salmon.[22] Many other chemicals affect fish behaviour directly. As salmonid fishes of different kinds are extremely important economically, the effect of pesticides has been investigated in depth for many more of these. Both sexual behaviour and homing (see Chapter 5) have been shown to be influenced by a host of the industrial chemicals we use in agriculture and forestry. Interestingly, behaviour was also influenced by cypermethrin, which is used to protect salmons from copepod salmon lice in the fish-breeding industry.

Another good example is 4-nonylphenol (4-NP), which is used as a ubiquitous surfactant both in industry and in sewage treatment plants. Today, this chemical is found in more or less every body of water over the globe. When scientists exposed social fish species to 4-NP concentrations equal to that found in nature, the effects were quite drastic. The fish stopped responding to pheromones mediating schooling and instead displayed opposite behaviours. The 4-NP pollution evidently has a direct effect on behaviour vital to both predator avoidance and feeding.[23]

Looking at all the numerous chemicals we produce and the different ways they add to nature's natural chemical diversity, it's clear that fish and other water animals suffer heavily from them. One way is through the direct and indirect effects on their olfactory life.

Sometimes the pollutants just seem to destroy the ability to smell, sometimes they affect smell-directed behaviour indirectly, for instance via the hormonal pathways.

Human smell

Let's return to the year 1022 and consider our own odour. As you will see in Chapter 2, one of the world's largest industries profits from the fact that we believe we smell inherently bad. Even though perfumes and perfumers existed in India, Egypt and Mesopotamia thousands of years ago, their use only really took off in Europe in the 1700s, with King Louis XV and Madame de Pompadour in France. They led a fashion trend in scents that everyone wanted to emulate. But further back in 1022, the fellow humans you encountered would still emit a more or less natural fragrance.

Another habit that has strongly affected our body odour is the practice of taking frequent baths and showers. These cleansing rituals also took off in the 1700s, as water started to be considered healthy even in the cities. Bathing and the use of soap changed the microflora of our bodies and thereby also our smell.

This is why, in the Anthropocene, we smell both less and differently compared to other epochs. By washing ourselves regularly we diminish our bodily odours, and, by applying foreign, smelly substances, we drastically change the smells we do give off. Often, these foreign formulations also contain deodorants that kill off microbes on our skin to further augment the change in odour Gestalt.

This change probably also means that we gain less information about each other. As you will discover in Chapter 2 and also from examples from other species throughout this book, there is a lot of information hidden in the smells we emit. A substantial part of this is very likely lost in our attempts to camouflage our true smelly self.

Our own sense of smell and the Anthropocene

Whilst we constantly strive to hide our smells, in the process we may be also losing our ability to smell. Our modern world may be partly to blame for olfactory dysfunction. While it's generally accepted that poor air quality can lead to debilitating respiratory and heart conditions, olfactory impairments resulting from pollution are only now beginning to attract more attention.[24]

There may be a link between air pollution and the risk of mental health issues or neurological disorders – including Parkinson's and Alzheimer's disease. While poor air quality is not a definitive cause of any of these neurological conditions, studies do suggest that people might be more at risk of developing them if they live or work in areas that are highly polluted, particularly if that pollution is caused by sooty particles.[25]

And the connection to our sense of smell? In both Parkinson's and Alzheimer's disease, anosmia (acute loss of smell) is often one of the signs that someone could be suffering from one or other of these diseases – or will probably suffer from one in the future. Anosmia is also often linked to cases of depression or bipolar disorder (see Chapter 2).

This area still needs to be investigated, but it's quite possible that there is a link to our olfactory sensory nerves and the flow of cerebrospinal fluid – the cushioning liquid around our brain and spinal cord that is also responsible for getting rid of waste products from our brain cells. There is some evidence that this fluid leaves our body through our nasal passages – as well as through the lymphatic system. If our olfactory sensory nerves or pathways are damaged in any way – even from air pollution – this could have a knock-on effect and trigger neurological conditions. However, the science is not yet conclusive, and more research is underway.

Disease and smell

For thousands of years we have been cohabiting with our pets and domestic animals. The first companions were probably dogs, and then came pigs, cows, horses and many other animals. In 1022 many people shared a single room together with their family members and their animals. This meant that we also started sharing microbes with the animals, which was the start of many diseases.

As humans multiplied and populations got denser, we created an optimum environment for these diseases to spread, and some of them had a direct effect on our sense of smell. The most recent example being the Covid-19 pandemic. In this case, the virus is thought to have spread in Chinese animal markets, where live, feral animals are handled and transferred between very many people at close quarters. The virus had plenty of opportunities to jump over to the numerous humans milling around – and start spreading around the world.

One of the symptoms that occurs in most Covid-19 patients is a total loss of both smell and taste. Whether taste really has gone needs to be revisited, however, as what most people think is taste is actually retronasal smelling. Anyhow, the research on the loss of smell in Covid-19 patients is going on both at the peripheral level – the nose – and at the central level – in the brain. So far, some results show that specific support cells around the smell neurons in the nose seem to be affected. Intense research is also investigating effects in the olfactory bulb of diseased Covid-19 victims.[26]

In a few years, we will probably know the exact mechanism with which this virus shuts down the capacity to smell in affected humans. Whatever the cause, it is clear that it is the practice of humans living together with our animals that has caused the transfer of harmful microbes between the species. We should consider this fact in our relationship with animals, especially with wild ones, but also in our modern practice of keeping domestic animals. The tighter we squeeze them in, the easier disease will spread. The frequent use of antibiotics to allow the dense packing of animals in industrial farming is another question entirely, and absolutely vital to human survival. But that belongs in another book.

Chapter 2

Human Smelling and Smells

There are many aspects to human smells. We pick them up and give them off. They attract and repulse, elicit disgust or desire, and even warn us of danger and disease. The sense of smell, or olfaction, helps us to perceive and make sense of the chemical world around us and is in many ways crucial for us to function safely, healthily and happily as humans. And yet, we tend to disregard this fifth sense or see it as a relic from primeval times.

Smell is not the sense that comes to mind when we want to differentiate ourselves from other living creatures. That honour would go to sight and hearing, closely followed by touch and taste. In some quarters, even our so-called sixth sense – a supposed special intuition or power of perception – might get more traction than smell.

Is smell too primitive a sense for us civilized human beings? Would we rather focus on the sharp lines that separate us from animals than accept any that might be blurred? For many, recognizing smell as an important sense may bring us too close to animal behaviour for our own comfort.

However, judging by the fortunes we spend each year on concoctions to rid us of smells – and cover us in scent

– sensing and secreting the right smell is important to so many of us. So important it's become a multi-billion-euro industry. You may be aware of the individual scents you purchase separately, such as perfumes and fresheners, but you are mostly unaware of the big businesses working constantly in the background, subliminally scenting nearly all consumer products and environments.

The mall you shop in is typically scented, often with a branded scent. The clothes you buy at this mall are almost certainly mostly scented, too, again usually with a branded scent. Even if you don't buy anything and just stop for a coffee, it's more likely that the smell of the coffee shop that tempted you in wasn't actually the smell of freshly ground or brewed coffee as you thought, but rather a branded coffee scent emitted from a machine under the counter.

This massive industry, its figurehead a conglomerate known as International Flavours and Fragrances (IFF),[1] a company that sells scents not in thin bottles but in tanker trucks, has not arisen purely for our pleasure and to rid us of our shame. Surely, there's more to our scent than human vanity.

Indeed, why would our nose and nostrils feature so prominently on our faces if our sense of smell weren't crucial for survival? Well, in some situations, it is. Olfaction is a constantly analysing sense. It monitors the quality of potential food items, it scans the environment for potential hazards, but it also provides the fine nuances of pleasure from eating a strawberry, imbibing your favourite Malbec and snuggling into your loved one's armpit.

The analytical side of our sense of smell becomes very clear when comparing it to our sense of taste. Taste is built on five types of rudimentary taste sensations (salty, sour, bitter, sweet and umami) and is basically there to get noxious stuff out of your mouth as quickly as possible – more or less as a reflex. While smell, with its around 400 receptor types, analyses chemical details to allow approach to good food, drink and other items of value or, conversely, to trigger approach avoidance to bad stuff.

Smell provides us with information that is important for our nutrition, our safety and our quality of life. Losing the ability to detect smell is a pretty big deal. Its absence can be accompanied by mental health problems – as people can no longer enjoy food, drink or life in general. They often worry incessantly about their personal hygiene and miss out on the sensual scent of their loved ones.

You lose it, you lose out.

A terrible loss

This sense that most people take for granted was thrown afresh into the spotlight in the UK on 18 May 2020, when the nation's four chief medical officers issued a joint statement: "From today, all individuals should self-isolate if they develop a new continuous cough or fever or anosmia," it was announced. Fortunately for those members of the public who had never heard of "anosmia", the scientists followed up with an explanation: "Anosmia is the loss or a change in your normal sense of smell."[2] (It's worth mentioning that the physicians were not perfectly accurate in their statement. While a loss of the sense of smell is indeed termed "anosmia", a change in the sense of smell is termed "parosmia".)

The long-awaited announcement came as evidence had been piling up that a sudden loss of smell could be an early warning sign of the novel coronavirus disease 2019 (Covid-19) caused by severe acute respiratory syndrome coronavirus 2 (SARS-CoV-2). The disease's olfactory symptoms could serve as a potential biomarker. And anything that could give a patient an edge in beating this disease was welcome.

Initially, the evidence seemed purely anecdotal that anosmia was the hallmark of Covid-19. As the disease spread and the numbers of infections went up, so too did the reports of anosmia, also within the medical community. This led chemosensory scientists from all over the world to investigate and share their understanding of this specific type of anosmia, often with a considerably more open-source approach to science than usual, sharing data and research results in real time. Undeniably, this method has upsides, but there are also pitfalls.

In the frenzy to uncover and share new details on the disease, scientists have been releasing so-called preprints, research papers that have not been peer-reviewed, meaning that they had not undergone the usual scrutiny. In the circumstances – an unprecedented pandemic – this was a good thing for the scientists. For the general public, it wasn't always. Journalists would seize on the information and publish it with click-baiting headlines and draw conclusions that often showed a lack of real understanding of the science.

Some preprints do hint at answers, although many mysteries about the disease remain. What seems likely – at least at the time when this book was going to press – is that anosmia or parosmia can be common early neurological symptoms, and in some cases the only symptoms, of Covid-19. To date, most scientists appear to agree that SARS-CoV-2 uses the angiotensin-converting enzyme 2 (ACE2) receptor and the transmembrane serine protease 2 (TMPRSS2) to latch on and infect cells via Covid-19's spike protein.

Both ACE2 and TMPRSS2 are abundant in the nose, throat, and upper bronchial airways, and are particularly plentiful in the respiratory epithelium and in the supporting cells of the olfactory sensory epithelium inside the nose. Even though ACE2 is in fact rare in the olfactory receptor neurons themselves, the supporting cells of the nasal epithelium could therefore provide the cellular opening for the coronavirus.[3]

This could explain why the sense of smell is affected in the initial stages of the disease, without the patient having symptoms of a stuffed nose or breathing difficulties. And why also older people are more susceptible to the disease – as they have more of ACE2 receptors than younger generations.

There is evidence that the virus can migrate into the central nervous system via the nose and olfactory bulbs, as well as by other routes. As noted above, Covid-19 doesn't seem to target or invade the sensory neurons directly,[4] meaning that the anosmia is not generally caused by damage to the central nervous system, and the sense of smell does return. Some patients, however, have still not recovered their sense of smell even months after recovery from the disease – there are case studies where patients have been left with a 10–20 per cent reduction in their olfactory sense, which suggests that either the olfactory sensory neurons have been permanently damaged, or that the central nervous system is in fact involved. Brain scans that have uncovered signs of the disease in a patient's brain would support this theory. Or, it could simply be that the olfactory sensory neurons need longer to regenerate after such a disease. Again, it's all so very mysterious.

At first sight, anosmia in Covid-19 patients may seem a trivial neurological symptom given the other possible outcomes of the disease, but further research in this area may provide insights into the progression of the disease, how it operates and its likely outcome. And that kind of understanding should lead to better treatments, not only for Covid-19 but also perhaps for anosmia. A tantalizing case has recently been reported in Israel where a woman who was totally anosmic for 12 years following a viral infection regained olfaction after Covid-19 infection.[5] This case is not sufficiently well documented to build on, but it does further add to the mysteriousness of the Covid-19-olfaction link.

Anosmia is not just a potential symptom of Covid-19, and it's never actually trivial. It often accompanies or follows many types of viruses, colds or respiratory diseases. Head trauma can trigger it, as can allergies, radiation treatment or even a drug addiction to cocaine. Not being able to smell is often a signal that something else is wrong. A sinus infection may inflame the tissues in the nose and damage the sensory cells, while head trauma can damage the olfactory nerve fibres that lead to the brain.

Both Parkinson's and Alzheimer's disease affect olfactory acuity and smell tests are used for early diagnosis (see Chapter 14). Often, we can therefore pinpoint the trigger for anosmia, but may only speculate at what is actually going on in our olfactory organs and brains. The reason is simple: our nose remains a bit of an enigma in the scientific world. It is complex and sensitive.

How sensitive are we?

Can scientists put a number on how sensitive our noses are? They have certainly tried. In the 1920s, rough calculations – not much more than guesstimates really – put the number of odours that humans can detect at 10,000. That remained the number cited in a lot of literature, but it's a mystery exactly how such a nice round number was arrived at, as Avery Gilbert discovered when he tried to track down the source for his book *What the Nose Knows: The Science of Scent in Everyday Life.*[6] Scientifically speaking, he concluded, the data behind it was probably dodgy.

Almost a century later, in 2014, researchers at the Rockefeller University in the US upped that estimate considerably. Neurobiologists calculated that the actual number of discernible distinct odours was probably closer to one trillion.[7] That's quite a jump.

In the study, volunteers (none working in a profession that required strong odour detection skills) were asked to sniff a number of mixtures of unfamiliar odours made up of a combination of 10, 20 or 30 odour molecules from a resource bank of 128. The volunteers had only to spot the odd one out in groups of threes. Extrapolating from these results, the study suggests that an average human – untrained in the art of scents – would be able to discern at least a trillion different smells. The least sensitive among us may still make out almost a hundred million different odours.

Not so fast. Other scientists dispute the "at-least-a-trillion" claim, arguing that the mathematical modelling used in the study is fragile.[8] The true number may always remain elusive, but we do know that we are able to pick up on many of the smells out there, and that, whatever the numbers, this remarkable sense is coded in our genes.

In our genes

With every breath we take, we sniff and smell the volatile odour molecules in the air. Roughly one to three per cent of our genes are dedicated to sniffing them out, recognizing them and triggering a reaction. For this insight, we have two Nobel laureates to thank, Richard Axel and Linda Buck. They were jointly awarded the Nobel Prize in Physiology or Medicine in 2004 for their pioneering work on unravelling how scents are detected in the nose and converted into signals in the brain. Even though their work was performed with mice and rats, it provided the basis for our first insight into human olfactory receptors.

What's known is that the odour molecules pass over the moist olfactory epithelium, located high up in our nasal passages on the roof of the nasal cavity, and over the millions (estimates range between six and 12 million) of odorant receptor cells that are embedded in the lining. Tiny cilia on each receptor cell pick up the odorants as they enter the nose. Each cell in this lining contains just one type of olfactory receptor (OR), from around 350–400 different OR types, and is specialized in sensing a limited number of odorant molecules.

The odorants bind to these receptor proteins, which activate the receptor neurons, which in turn transmit electrical signals, or neural impulses, to the olfactory bulb in the brain, bypassing the thalamus (which processes sound and vision sensory signals) en route to the limbic system, and thereby giving this sense a deep connection to our emotions. Our sense of smell lives in our limbic system, that part of the brain responsible for our emotions, mood, behaviour and memory. This instant processing may also be a reason why we can't always put a name to the smell, but more on that later.

But what is the purpose of these messages? What do they trigger in humans?

Human pheromones: fact or fiction?

Research into the different aspects of smells and smelling in humans is enormously contentious, particularly when it concerns studies that look into pheromones, those chemical substances that when released by one individual of a species trigger certain behaviours or processes in other members of the same species.

The word "pheromone" was coined in 1959 from a need to describe chemical substances that are used to communicate between individuals of the same species. The German biochemist Peter Karlson (1918–2001) and the Swiss entomologist Martin Lüscher (1917–79) created the name from the Greek *pherein* (transfer) and *hormone* (excite). The two were not the first to think about such communication – the ancient Greeks themselves had speculated that secretions in a female dog in heat attracted male dogs.

As you will discover in the following chapters, pheromones can trigger sexual attraction, but also other reactions that are essential for the survival of many species, including aggression, maternal instincts, as well as alarm and territorial behaviour.

Could such a whiff trigger survival responses in humans? Are human pheromones a fact, fraud or just complete fiction? Many scientists have pushed back hard against the concept of human pheromones, but let's look at what we do know.

The missing organ?

In other mammals, pheromones are detected primarily in the vomeronasal organ (VNO), a separate specialized olfactory structure in the nasal septum. Dogs, pigs, horses and mice all have sophisticated VNOs. And humans? It's disputed as to whether humans have this organ at all. Its existence would be an argument supporting the existence of human pheromones. Its absence would be a key argument against. But… not necessarily.

In most humans, it does indeed seem to be absent, but some scientists argue that even in its absence, the human olfactory system recognizes and reacts to pheromones. Others remain sceptical and argue that even in its presence, while chemical communication occurs in humans, we do not depend on pheromones for survival the way that other living creatures do.[9] So, who's right?

Intriguingly, studies have shown that human embryos start off with early signs of developing a VNO, but then seem to give up along the way and then lose most of it before birth. In a similar way, we possess both gills and tails as foetuses. The ontogeny mirrors the phylogeny. Endoscopic probing has found that adults are often left with a simplified version – a type of orphan VNO pit without sensory neurons or nerve fibres – if we are left with anything at all.

Nevertheless, despite the seemingly bare-boned version of a VNO, experiments have demonstrated that humans will react to interspecific chemosensory cues, but that such cues are processed by the main olfactory system.[10]

There is clearly a lot of conflicting information and evidence out there, as well as commercial interests, so more independent work is needed before anyone can conclusively claim that they have found the definitive answer. But whatever the anatomical structure of our noses, studies suggest that certain odours can trigger behavioural responses in humans. Are they pheromones, though?

Sex-specific triggers

Researchers have worked hard to identify pheromonal triggers in humans, particularly ones that may affect females and males differently, and, more interestingly perhaps, ones that induce a male- or female-specific physiological response involving our reproductive systems.

Such studies invariably monitor cerebral blood flow while the subjects sniff an estrogen-like substance (similar to the one excreted in female urine) or get a good whiff of the compound androstenone (the key mating pheromone for pigs and also the testosterone derivative secreted in human sweat, most notably in male armpit sweat) or its close relative, androstadienone. Some researchers claim that males and females will react differently to these compounds[11] – and that their reaction involves our hypothalamus, that tiny region of the brain tied to the regulation of hormone secretion (and therefore sexual reproduction) and homeostasis – which is necessary for maintaining our body in a state of equilibrium.

Studies have demonstrated that this area reacts to estrogen-like compounds in males, but in females it lights up more when the scent comes from androstenone. Other highly contentious studies even go so far as to suggest that the same putative human pheromones could trigger different sexual arousal responses in lesbian women compared to heterosexual women, and also in homosexuals compared to heterosexual males. Scans showed that estrogen-like substances triggered the same brain activity in the hypothalamus in lesbians as they did in heterosexual males. And that homosexuals reacted the same way to androstenone as females did. Evidence, perhaps, that the hypothalamus reacts according to sexual orientation?[12 13]

When you consider that all these studies involve very small numbers of subjects – and also use a far more concentrated version of the odour than you might encounter in real life – these results lose some of their punch, become problematic and ultimately are not entirely convincing or conclusive.

Weak experiments are not evidence that smell is not important to humans – or that human pheromones don't exist. As the British evolutionary biologist Tristram Wyatt points out, we should be sceptical of such research, but should also be open to the option that pheromones do exist. We may find better answers to the elusive pheromones by rigorously studying the very obvious changes to our sebaceous glands and body odour as we move into puberty and adulthood (think of the typical stink of a teenage boy's bedroom). And also, in the secretions from the areola glands around female nipples produced by all lactating mothers.[14]

Baby needs and heads

Could mothers produce the most important human pheromone? An odour that is secreted by the skin around the nipples of lactating mothers seems to trigger a survival instinct in newborn babies – that suckling reflex. Studies have determined that babies, and not just the mother's own, will go into suckle mode when they pick up the scent that is secreted in the nipple area by any breast-feeding mothers. The fact that the reaction was observed to take place irrespective of whether the mammary scent comes from the baby's own mother or not would indicate that this could be a general pheromone.

But then again... because of evidence of a certain specificity, this odour may not meet the definition of a classic pheromone. Scientists argue that babies might learn their mother's own olfactory signature while being breast-fed and will then be able to recognize her by that unique scent alone.[15] Mothers recognize their babies by their unique scent, too. For babies, the mother's scent is so powerful that it alone can soothe a distressed baby – and prime a hungry one for feeding. Just the scent of breast milk can even calm a pre-term baby during and after heel-stick procedures. Pheromone or not, it's certainly powerful.

Is it all in our heads? The baby's head and your own? Nuzzling a baby's head has got to be one of the most satisfying feelings that there is. An "I could eat you up" might cross your lips as you snuggle up close ("*Du bist zum Fressen süss*" if you're German, "*θα σε φάω*" if you're Greek). Fortunately, unless you're a Greek god (Kronos), you probably couldn't actually bring yourself to do it.

What's behind this feeling? According to research led by Johan Lundström at the renowned Monell Center, the institute dedicated to interdisciplinary basic research on the senses of taste and smell, the smell of a baby's head triggers the reward circuits in a mother's brain (but not in a woman who doesn't have children). The smell elicits a physiological response very similar to the one experienced by hungry people when presented with a delicious meal. It's thought to be part of an evolutionary bonding mechanism.[16] It's probably more about the desire to be as close – and protective – to one's child, than to actual cannibalistic thoughts, fortunately. Scent could be a powerful tool in the human bonding mechanism.

A unique "baby" odour may be key to this mechanism. A recent study, performed in Japan and published in 2019, concluded that babies have a uniquely recognizable scent.[17] Researchers were able to take odour samples from the heads of babies and the mothers' amniotic fluid. They identified aldehydes, carbonic oxides and hydrocarbons among the 37 volatile odour components in all the samples. Sixty-two volunteers were first asked to smell one sample and were then later asked to identify the scent from a selection of four. Females were better at identifying the right scent, over 70 per cent correctly chose the baby scent from the others, but both males and females were good at recognizing the newborn scent from a specific baby that was only two to four days old.

The baby scents were definitely more distinct from each other, but most noticeably distinct from the amniotic fluid. This was determined using two-dimensional gas chromatography coupled with mass spectrometry. Although the study was performed with the scent from only five babies – understandably there are ethical aspects that render this kind of study more sensitive and trickier to perform than others – the findings seem to be based on solid methodology.

Does scent play a role even before birth? Let's look closer at that amniotic fluid.

The earliest scents

Although the scent of a mother is usually the first smell a baby will experience in the outside world, it's not the very first odour it will ever come across. The first is the scent of the mother's amniotic fluid. By the fifth month of development in the womb the baby is swallowing, sucking and digesting amniotic fluid.

Scientists in France have determined that a mother's diet will influence the scent in this fluid, and then this chemosensory information in the fluid can go on to affect a newborn's preference for the scent of certain foods.[18] The study showed that when expectant mothers consumed anise-flavoured food or drink during the later stages of a pregnancy, their babies would turn their head towards the smell in anticipation of feeding.

Similar studies also confirm that aversion or attraction to certain scents is learned already in the womb. We can assume that the baby associates familiar odours and tastes with positive experiences, and then logically prefers the same smells and flavours after birth. So, what's the evolutionary explanation for this? Well, generally mother knows best. If the mother is eating a certain kind of food, she obviously likes it and survives on it. If the foetus can learn that these smells and tastes in the womb are good, after birth they will be drawn to the same smells and want to taste the flavours again in breast milk, and maybe also in their food after weaning. This is where our taste preferences in life begin.

What about fathers? Does the baby smell have a similar effect on males? Although my passion lies with pheromones and moths, after becoming a father I couldn't resist doing some research on human pheromones. I wanted to know if babies really did have a particular scent that adults pick up. And if they did, whether it could influence our behaviour towards them.

Who's the better detector?

First, we partnered up with researchers at the Munich Institute of Medical Psychology, who are experienced in working with human aromas. In our study, we worked with twenty-four Swedish newborn babies between the ages of one and four weeks, and twenty-four older children, between the ages of two and four-years-old. The babies and children were bathed in unscented soap and then put to bed wearing special research T-shirts (such T-shirts, you'll discover, will become a bit of a trope in human pheromone research) and a clean little cotton bonnet.

These scented garments were then used to determine whether the mothers, the fathers or unrelated random childless men or women (twenty-four people in total) could pick out the one worn by a newborn baby from a sample of three (one newborn, one child, one unused). Each person was asked to perform the task with twenty-four groups of three.

Surprisingly for all involved, it wasn't the mothers or the childless women who scored highly, but the fathers. They were clearly better at telling apart the newborn scents from the odour of young children. Also, surprisingly, the women generally preferred the fresh, unworn garments to all the others.

Interestingly, using a gas chromatograph we were able to identify additional separate components in the scent that the babies secreted into their shirts and bonnets. These components were absent or much lower in concentration in the garments of the older children. One reason for this is very likely that in a newborn baby the sebaceous glands in the skin are almost as active as it is in adults for up to a week after birth. During childbirth, certain substances from the mother's body are passed on to her baby through the placenta. These act as a temporary stimulus for the production of skin secretion. In older children, this process of secretion occurs less frequently and only really kicks in again during puberty.

The conclusion of the study was that men generally have a better capacity to distinguish infants from toddlers. When we asked how the men would characterize the infant odour, they described it as "soothing", "calming", and "sweet". All positive descriptions and, in general, words that indicate the somewhat tranquilizing effect the scent seems to have. We can only speculate why men would have evolved such a capacity.[19]

As always, our genetic set-up reflects our genes' survival rates tens of thousands of years ago, so maybe the aggressive, male hunters, returning to the cave, showed some higher forbearance with the noisy little infants due to their wonderful smell. Who knows? Anyhow, after the study was mentioned in the press, I was interviewed by the BBC. Their last question was: "Will it now be possible to synthesize infant odour and spray it over football stadiums to calm down the hooligans?"

The smell of fear

Let's stay on the topic of aggressive behaviour for a moment and what it might trigger – fear. Do you think you could detect anxiety or fear in someone just by their smell alone? Numerous researchers have tried to determine whether "the smell of fear" might be expressed in sweat, notably psychologist Denice Chen.[20] Several involve collecting armpit sweat after exposing donors to clips from comedy or horror films and subsequently exposing other subjects to the pads of sweat and gauging their reactions. In one study, for instance, the volunteers were able to differentiate between the "happy" and "scared" male scents more often than by chance, suggesting that humans may give off anxiety and fear-related chemosignals after all. More intriguingly perhaps, we seem to be able to pick up on them, too.

Sweat is not the only bodily secretion investigated as a potential chemosignal carrier that can affect aggressive feelings. A different study[21] found that sniffing emotional tears collected from women then reduced testosterone in men. The authors speculated that the chemical signal in tears serves as a sort of chemical "stop" sign, reducing both aggression and sexual behaviour in conspecifics.

That sniffing tears reduces testosterone was independently replicated in humans,[22] but more interestingly, it was then replicated in mice,[23] and showed that sniffing juvenile tears reduces adult male aggression. We label this as interesting because most human chemosignalling studies entail an effort to replicate in humans effects first seen in rodent, yet here an effect first seen in humans was then replicated in rodents.

A certain chemistry?

Now, back to those noisy infants. Before you get to snuffle a baby's head and inhale their soothing odour, a whiff or two of an adult's armpit might be on the cards. But this time, we're not interested in the smell of fear, but the smell that could trigger lust.

Pheromone scientists seem to focus a lot on armpits in their studies. Which is hardly surprising when you consider that armpit smell develops in line with puberty progression. Located in our armpits – and pubic regions – are our apocrine glands, a special type of sweat gland. Unlike our eccrine glands, which are found all over our bodies and release a clear, watery, non-odorous and salty sweat that helps us to regulate our body temperature, our apocrine glands give off a fatty substance into our hair follicles that has the potential to be quite smelly. (The off-putting type smell is not inherent but caused by bacteria breaking down the fatty substance when it reaches the surface of the skin, hence the reason why deodorants work under our armpits). Does the link to puberty suggest that the smell is there for a reason?

In my university classes on the senses, I let students take part in an experiment to show them how good we are at distinguishing men and women by smell alone. I ask them all to take a shower with a non-scented soap the evening before my lecture and to put on no deodorant or perfume in the morning. Before the lecture, I distribute gauze pads to everyone and ask them all to wear these under their armpits during my morning lecture. Afterwards, each person takes off their pads, puts them in a jar and marks the jar with an anonymized number that only reveals the sex to me. Next, everyone smells through the jars and rates if they think the smell in each jar came from a man or from a woman. The result was always highly significant, pinpointing the sex of the smell donors correctly in about 80 per cent of the cases. There were, however, also examples of opposite ranking throughout.

At the end, the students are asked to rank if the smell is strong or weak and if it is pleasant or not. It turned out that all strong and unpleasant odours were ranked as male, while the weak and more pleasant ones were ranked as female. And this was obviously often true. Body odour is complex, but we basically have our apocrine axillary glands to thank for the strongest of human smells, and androstenone and androstadienone. These related steroids increase during puberty in both males and females but are more prominent in men.

Apparently, our genes determine whether we perceive the odour from the two as pleasant or unpleasant.[24]

As well as perhaps helping us to tell males and females apart, does our sense of smell help us to choose a mate?

Genetics and an immune boost

Going by the apocryphal story about a horny Napoleon Bonaparte – the one where he allegedly writes to his first wife, Josephine, from the battlefields telling her "Don't wash. I'm coming home!" – body odour can have seductive qualities. Is this simply a strange personal foible of Napoleon, or is it something that is evident across the entire human species? And what would be the point of sniffing our potential mates if it were?

One idea that scientists have pursued is that a person's body odour can tell us a lot about someone's immune system. As it could pay to choose a mate with an immune system different from our own, is it possible that we instinctively know whether someone would make a good mate or not simply by paying close attention to their scent? The idea being that any offspring from such a partnership would likely be more resilient.

To understand how this might work in practice, scientists have looked at what makes us resilient. All vertebrates have a collection of proteins that are carried on the surface of every cell. This group of cell-surface proteins is referred to as the major histocompatibility complex (MHC) and it helps regulate our immune system. Humans have the human leukocyte antigen (HLA), which is encoded in our genes by the MHC. Could this be the measure of the quality of a potential mate?

In one study[25] – also involving well-worn T-shirts – women sniffed shirts worn by men and picked the one worn by a man they would prefer to socialize with. They tended to select shirts from men with MHC genes that differed from their own. Women on birth-control pills, however, showed the opposite effect and were drawn to MHCs similar to theirs. A sign perhaps, the scientists speculate, that as the pill puts the body into a hormonal state in part similar to pregnancy, you would choose to have similar relatives around who you would expect to be supportive. There's just one caveat, however. So far, such results have not been replicated elsewhere.

Another study determined that women tend to prefer a perfume that had the scent similar to their own MHC immune proteins.[26] The researchers speculate that this would indicate that while women prefer potential partners to smell of MHCs that are different from theirs, they prefer the smell of their own MHCs on themselves. The reason for this preference? Maybe they sensed it would boost the smell of their own immune system. Again, as there were only a limited number of subjects involved in the experiment, more research is needed before you could say this conclusion is definitive.

In all of these studies, scientists looked at how we react to extremely close contact with volatile putative pheromones (the sniffing of T-shirts is carried out up close). It seems that long-range ones haven't been researched enough yet, are not in our repertoire, and/or possibly don't have the capacity to change our preferences or hormonal states.

CHAPTER 2

Women: in and out of synch

Talking of hormonal states, is it possible that women can influence each other's menstrual cycle by the pheromones they secrete?

One of the most controversial studies on female ovulation was hailed as a breakthrough in pheromone research when it was first published in *Nature* in the 1970s.[27] The Harvard paper, titled 'Menstrual Synchrony and Suppression', had tracked 135 female college students living in the same dorm and observed that their start dates for their periods showed synchronization over time. The psychologist Martha McClintock, who led the study, concluded that the observed synchronization of menstrual cycles among friends and roommates must be due to pheromones and that there is "some interpersonal physiological process which affects the menstrual cycle". At the very least, it confirmed what many women describe anecdotally: namely that when women live together, their cycles will invariably seem to synch over time.

In the meantime, few studies have reliably and unequivocally replicated the results. In one instance, another study involving the dabbing of female armpit sweat on the upper lip of females observed that their menstrual cycles were affected by the odour – they shifted in synch to the sweat donor's cycle.[28] However, the subjects were few in number.

One carefully controlled study in 1998 does stand out, even though the number of participants was also low.

Co-authored by McClintock and Kathleen Stern,[29] it concluded that pheromones do have an impact both on the timing of ovulation and the length of the menstrual cycle. The researchers had studied the effects of the odours from armpit sweat from other women over four months. For two months, ten subjects were exposed to odours from the follicular phase of the menstrual cycle, and for two months ten others sniffed the odours from the later stage in the cycle. Intriguingly, the "early" subject's menstrual cycles were shortened – by an average of 1.7 days a month but sometimes by as much as 14 days. However, the cycles were lengthened when exposed to "late" scents – by an average of 1.4 days a month but also up to 12 days. The scents came from a pool of nine female sweat donors. The chemical molecules involved were not identified.

Other scientists have strongly challenged what has been dubbed the McClintock effect in its entirety. The observed synchronization is put down to pure chance or even flawed science.[30] As most women have cycles of different lengths, the odds are that at some point during their cycles there will be some convergence in the cycles of women who live together.

Whatever the claims, what we should perhaps be asking more loudly is what the evolutionary value to this putative pheromonal trigger could be. One theory is that it is the female way to bond and co-operate against the male. Or does it reduce or increase the competition for the male partner? Menstrual synchronicity remains a much debated issue and a hotbed of controversy.

At the heart: approach-avoidance

As mentioned at the very beginning of this chapter, one of the main functions of the olfactory system is to monitor constantly our chemical surroundings and to use these cues to warn us of dangerous situations. Each time we inhale we get information that allows us to make informed decisions about where to go and where not to, what to eat and what not to, and, to some extent, who to befriend and whom not to! It turns out that the most salient signals seem to be those connected to negative situations: the smell of someone who threw up (obviously some bad food around or someone who is sick), the smell of smoke and burning (obviously danger of getting burned) or the smell of spoiled food or drink.

In addition to these more or less innate signals, we also very quickly learn to associate a specific odour with something bad. Eating something and then getting really sick is the most common experience. And it's one that often results in a lifelong repulsion to this odour, be it melon, meatballs or mascarpone. Such associations can also trigger anxiety.

Johan Lundström at Karolinska in Stockholm has performed elegant experiments to try to understand how our brain builds up these negative associations.[31] Pairing specific, previously neutral odours with electric shock, Lundström could make his test persons build up negative impressions of these odours. Using odour molecules that are indistinguishable from each other, the researchers tested whether the subjects could tell them apart if one of the two was paired with an electric shock, which simulated an unpleasant experience. At first, the test persons were unable to discriminate between the odours but became better at identifying the ones associated with the shocks, even in lower concentrations. After eight weeks, the subjects were invited back and were again asked to discriminate between the odours. Intriguingly, the test persons had not retained any sensitivity to the "shocking" odour, however. It's not entirely clear why.

Interestingly, another study determined that physicochemical properties might be to blame for our long-term aversion to certain smells.[32] More odour molecules that are more complex structurally could result in higher trigeminality (the level of irritating components in an odour – it's what gives chilli that kick when you taste it), the researchers suggest. The study determined that the more complex the odour, the more likely it was that it would impact habituation – the subjects would always rate such odours as unpleasant. In other words, the neurons detecting these warning odours showed very little or no habituation. They didn't get used to a bad odour being around and didn't stop responding to it, which is what happens to most positive or neutral odours.

Personally, I experienced this non-habituation some 20 years ago when I drove up to the very north of Sweden with my then three-year-old son to take part in the annual reindeer cull. Going home, after a pizza of considerable size, my son threw up in the back seat. For the next 1,000 kilometres, from Sveg to Lund, I smelled the bouquet of puked pizza with every breath I took.

So unpleasant

Many odours have an inherent value, valence or hedonic tone to the human nose. When my friend Noam Sobel, a renowned neuroscientist at the Weizmann Institute of Science in Israel, tried to measure how we categorize odours and connect this to the chemical characteristics of the molecules, the only significant parameter he could find was exactly this hedonic tone – whether it was nice or nasty, pleasant or unpleasant.

In quite a complex experiment, together with scientists at the Neuroscience Institute and Department of Psychology, University of California, Sobel and his research team, the Weizmann group, set about trying to identify the general principles by which our sense of smell is organized.[33] They started out with a database of 160 different odours that had been ranked by 150 perfume and smell experts according to a set of 146 characteristics. These included "sweetish", "smoky" or "musty". The team then analysed these data to isolate a single factor that best distinguished the odours from one another. Their conclusion: it was all down to hedonic tone – the perceived pleasantness ranking of an odour.

This ranking has "sweet" and "flowery" at one end of the scale, and "rancid" and "sickening" at the other. The researchers then performed the same type of statistical analysis on a database of chemicals, taking account of more than 1,500 characteristics of each chemical, to come up with one factor or number that best distinguished them from one another. And that turned out to be the hedonic tone again. This meant, they argue, that it is possible to predict how pleasant a smell will be to humans just from its molecular structure.

Intriguingly, the same experiments also showed that the odour receptors in our noses tend to be grouped together according to whether they react to and identify a pleasant smell or an unpleasant one. That doesn't mean that an individual's cultural environment or experience won't influence the perception of some smells, or the organization of cells in our nasal lining, but it does seem that there could be a kind of global consensus on the most pleasant or unpleasant types of smells.

Noam summarizes it all as follows:

"Our findings show that the way we perceive smells is at least partially hard-wired in the brain. Although there is a certain amount of flexibility, and our life experience certainly influences our perception of smell, a large part of our sense of whether an odour is pleasant or unpleasant is due to a real order in the physical world. Thus, we can now use chemistry to predict the perception of the smells of new substances."

It's worth noting here that studies indicate that young children do not appear to distinguish between unpleasant and pleasant smells as distinctly as adults tend to. They recognize that an odour is strong or weak but will generally not qualify it as nice or nasty.

In general, it has turned out to be extremely hard for us to categorize odours outside of this hedonic tone – and usually impossible for most people to describe an odour using universally recognized terms. With this realization, the Weizmann group shifted from trying to predict the odour terms associated with a structure, and instead try to predict the perceptual similarity or differences of any two odorants, regardless of "how" they smell. This led them to a metric, a number they can assign to any two odorant mixtures based on their structure that indeed reflects the odorant's similarity.[34]

The Weizmann group claims that this will provide the basis for digitizing smell, but only time will tell if they are indeed approaching this long-sought goal. Moreover, what further complicates our dream of a digital future, is the fact that we are notoriously poor at identifying and naming odours.

Inside our brains

The reason for this apparent deficiency may be linked to the way our brain processes odours and language. If you have any doubt that smell or smelling is important to humans, just think about how we use language related to this function. If we don't like something, we kick up a stink or turn our nose up at it. If we suspect something is not right, we smell a rat. But, when we think it's a good idea to rely on our instincts, we say we should follow our noses! Maybe this is a sign that we feel instinctively that this is one sense that we can trust.

And maybe we can trust it, but we generally definitely can't describe smells adequately. Compared to the language related to sight and vision, we do not have a particularly rich vocabulary for smells. Even when healthy volunteers have been asked to name a common odour, they will invariably have problems putting a name to it. Studies point to areas in the brain that are to blame.[35] We have two brain regions that seem to be activated during odour naming: the anterior temporal cortex and the orbitofrontal cortex. According to the scientists involved in the studies, these areas of the brain receive fairly unprocessed olfactory signals, which make it more difficult for us to engage the language-processing areas of our brain to be able to identify and put a name to the odours.

In evolutionary terms, language-processing came much later than odour-processing. This could be the reason why our everyday vocabulary for odours is poor. Gaining a real understanding into how our brains process and link up the everyday experiences of smells to our language remains a challenge.

This linguistic aspect to smells led me to venture into the realm of human olfaction again – and to look at the ways that some cultures describe odours.[36] Together with my colleague Asifa Majid, who is a specialist on the topic, we studied how Europeans compare to a Malaysian rainforest tribe, the Jahai, in how they describe odours. We found that Europeans, exemplified by the Dutch in this case, would, in general, characterize odours by using concrete descriptors. In other words, they compared the smell to something familiar, such as a banana, for instance. They needed longer to describe the odours, too. Conversely, the Jahai were much more prone to use abstract descriptors, such as "musty", and needed much less time to describe the odour.

At the same time, we looked at the facial expressions elicited by the different odours and found that the emotional response to the odours was the same, despite the linguistic differences. The responses were similar, but the Jahai had learned to describe the smells with a specific abstract vocabulary for smells. In earlier studies, Asifa and her collaborators had also shown that the Jahai are extraordinarily good at distinguishing smells and have several specific words to describe certain important odours. In contrast, while we have "red", "blue" and "green" for colours, we lack something similar for smells.

The Jahai have a word that describes a different stinging smell and one that specifically refers to a bloody, fishy, meaty type of smell, and would recall them as easily as we would use our words for colours. All of their terms usually describe something of great significance to a life in the rainforest. As far as I remember, one word describes the bloody smell that would attract a tiger, which, the Jahai people say, is much like the smell of crushed head lice.

No smell, no taste

In my own teaching, I regularly perform another simple experiment together with my students to reveal the importance of retronasal smell – the one that gives us the ability to perceive flavours as we eat. Blindfolded and provided with a nose clip to prevent nasal breathing, I ask a group of students to tell the difference between ketchup and mustard. No one can. Minus the clip, but still blindfolded, everyone easily can. The explanation? Ketchup and mustard have about the same level of sweetness, saltiness and sourness. The real difference in taste only appears from the retronasal sensation of tomato versus mustard notes.

This is again one of the reasons why anosmia is considered to be such a handicap. More or less all fine-tuning of sensations of food and drink disappears with anosmia and everything becomes more or less the same. No difference between Macallan and Ardbeg whiskies. No difference between the best Rioja and a vino tinto.

So, when the sommelier pours you a glass of wine to "taste", you really should sniff it first. And swirl it around your glass as well, before taking another sniff, and only then a sip. The more the aromas, or bouquet, from the wine are released, the better your olfactory system can pick them up. And the more you'll enjoy the taste – if it's a fine wine that is.

Emotions and memories

Finally, we come to probably the most enigmatic part of the olfactory puzzle. Why is it that one whiff of something can transport us back in time, to another place and to another precise feeling?

This experience is often dubbed the "Proustian memory", "Proustian phenomenon" or even "Proustian moment", after the author Marcel Proust's description of an overpowering childhood memory. His was evoked by the sweet scent of a madeleine biscuit soaked in linden tea, which he wrote about at some length in his tome *In Search of Lost Time*, published between 1913–1927. But actually, any scent could do it.

The key to this type of "involuntary memory" is its sudden appearance out of nowhere, set off by a long-forgotten smell. And it is connected to a strong emotion that was felt at the time that the scent and event was saved in our memory.

Does Proust's poetic licence stand up to scientific scrutiny? The reason why scents and smells evoke strong, emotional-laden memories is that this sense is nestled in our limbic system, on a direct path to our amygdala, that tiny area in the brain where emotions are triggered, and memories are stored. A memorable smell can take us by surprise and stop us in our tracks, simply because there's no immediate conscious processing involved. The olfactory information is only passed on to the hippocampus afterwards. It makes sense then that such a pathway could result in evocative memories – and strong emotions. People with post-traumatic stress disorder know this all too well. They often report that familiar smells from the time of a specific trauma can trigger painful memories and even a deep sense of fear.

No trivial sense

Due to its brief nature, this chapter can only provide a narrow overview of human olfaction. It hints at some of the secrets and mysteries that are hidden right under our very own noses. What's clear is that our sense of smell is no trivial matter. It evokes strong emotions, unlocks memories, and even helps diagnose disease. Being able to smell helps us to enjoy life and our love life to the full.

You can learn how chemosensory scientists are trying to put their knowledge to good use in Chapter 14. Used strategically, the right kind of sniffing may even help delay that dreaded cognitive decline in old age.

Chapter 3

Our Old Friend the Dog and Its Exquisite Nose

From a dog's perspective, the daily walk never gets boring. To us, the same old route looks pretty much the same every single time. We're usually taking a walk just for the exercise. To dogs, however, the walk is not about keeping fit – it's all about the sensations and staying on top of what's going on. Dogs experience a very different world to the one of dog-walkers. That's why you'll often see owners almost comically pulling, yanking and begging their dogs to keep moving. "There's nothing new here," they seem to be saying. For humans, that may be true.

It's another story for dogs. Their astonishing ability to sniff out the faintest of odours radically alters their perspective and view of the world. And it makes them stop and pee, again and again, turning a would-be simple ritual into a tug of war (although not usually with the best-trained dogs). They also want to become part of the scented story. By leaving their territorial mark along the way, they are making their own contribution to the tale for the next canine that comes by.

With their noses to the ground or up in the air, dogs become emerged in their own kind of detective work, recreating a story that's invisible to us. It's a story that is coded in scents and odours, in smells that pass us by. While we tend to focus directly on the visuals —noticing only what we can see right here, right now, in the moment – dogs will become totally immersed in the air and the surfaces to build a picture of events and circumstances. Dogs smell history. It's their exquisite nose that opens up a completely different worldview and gives them quite an enviable edge over us.

Sniffing is a source of mental stimulation for dogs. It allows them to interpret and navigate their environment. Using just their sense of smell, they can create a whole storyline from the past into the present, and even, in a way, into the future. A path may look empty to you, but a dog will sniff out what has been going on along that same route long before you got there. Or get a whiff of something that is still away in the distance or hidden from view. It can sense danger. Or prey. Have you ever noticed how a dog reacts to a cat before you've had a chance to spot it? You can put that down to their olfactory prowess, too.

Dogs are able to zoom in on smells in a way that we cannot, including in on those odours that have lost their pungency over time, on those smells that linger faintly on. Dogs don't mind how faint the smell becomes, or how faint it is at the outset. Adrenaline, for instance, that fight-or-flight hormone we secrete when we are stressed or anxious, is pretty much undetectable to humans. Dogs, however, can easily pick up on it. It's one of the reasons why dogs are the go-to animals for emotional support. They really do know when we're feeling anxious or fearful. This ability may also explain why dogs are often most interested in people who are afraid of them.

A sensitive structure?

Why do dogs become so much more distracted by odours and attracted to smells than we seem to be? For a start, their odour threshold is much lower than our own, so they notice smells more. The numbers differ, but researchers have roughly estimated their threshold for certain compounds to be somewhere between 1,000 to 10,000 times lower than our own.

Their lower limit of detectability for odours in the air – volatile organic compounds – is reportedly one part per trillion.[1] What is it about the dog's anatomy that has led to this extremely low threshold and exact olfactory acuity?

Is it their wet and shiny nose? A dog has sweat glands on its nose, and on the pads of their feet, which help them to regulate their temperature. (They don't sweat over their whole body as we do.) A humid atmosphere in, on and around the nose makes it easier for them to absorb scents. That's why dogs tend to lick their noses frequently – to increase moisture and up their sense of smell. A wet nose also helps dogs sense the direction of the winds, as a wet surface gets colder where the wind hits it. Compare this technique to the way we hold up a wet finger into the air to detect the direction of the wind. And, as smells travel with the wind, once dogs know the direction of the winds, they will know where the smell is coming from. But that's not the key to their sensitivity. How the nose is structured and what's happening on the inside is what really matters.

At first glance, a dog's nose may appear unimpressive. They have two nostrils, just as we do. But, unlike us, they can move and use each nostril independently. This also helps them determine the source of a scent, or the direction from which it is coming.

Along with the independently and individually movable nostrils, dogs have side slits, or nares, by each one, something we're obviously – and thankfully perhaps – missing. The function of these slits is to enhance the absorption of a scent, by helping dogs to expel breath in a more focused and efficient way. They inhale through the nostrils, and exhale through the slits, which causes a current of air to push new odour molecules into the next breath the dog takes.

Researchers who have studied the external aerodynamics of dog sniffing and scenting have discovered that dogs don't exhale air over the source of a scent. Instead, their slits divert the jets of airflow away from the source, to the side and behind. This avoids disrupting or contaminating the source of the odour and ensures that scents are pushed forward towards the nostrils. Dogs "scan" over the source of the smell, sniffing frequently, until they zoom in on the source and drive the odour molecules upwards inside the nose directly towards the smell mucosa or the olfactory epithelium.[2] No air enters or exits the olfactory area during expiration. This means that any odour molecules are not diluted or disturbed and remain efficiently exposed to the chemoreceptors of the epithelium throughout the complete respiratory cycle.

As a side note, heavy panting is thought to create more turbulence around the source of the scent, and could therefore be the reason why tired dogs, or dogs who are trying to keep cool, seem to have a reduced ability to detect scents. The same goes for dogs whose diet and microbiota are not the healthiest, or who are simply out of shape.[3] Frequent sniffing, however, allows dogs to take in more odour molecules with each breath than we ever possibly could.

A long, circuitous route

Once the odour molecules have entered the snout, what happens next? Inside, the dog's nose is ideally adapted to assess smells. Odour molecules are taken on a somewhat circuitous, scenic route. They have to pass over a sophisticated olfactory apparatus, where a moist epithelium tightly packed with olfactory sensory neurons – the cells responsible for detecting odours – lines and almost fills their nasal cavity.

What characterizes the dog's epithelium is its intricately folded scroll-like design over complex bony turbinate structures. These ethmoturbinates create an almost labyrinthine journey for each breath of odour molecules, which, equally importantly, results in a large total surface area to assess the molecules. On this surface, dogs have a higher concentration of olfactory receptors than humans. It's estimated they have at least fifty times as many scent receptors as we do. Studies have shown that bloodhounds that have been specifically bred for tracking have up to 300 times the number of odour-detecting cells as humans do.[4]

As Alexandra Horowitz, psychologist and author of *Inside of a Dog: What Dogs See, Smell, and Knows,* notes, humans have about five million olfactory cells while dogs have hundreds of millions, maybe even a billion cells.[5] This labyrinth and large surface area help dogs distinguish complex odours. Their olfactory system makes our own look rather stunted in comparison. And that is even more true of the vomeronasal organ.

The dog's vomeronasal organ, also known as the Jacobson's organ, is an additional site for odour detection. In humans, this organ has virtually disappeared over evolutionary time (see Chapter 2). Sitting directly above the roof of their mouth, it helps trap and discern specific scents, often of lower volatility than others. Licking increases the odour absorption in this organ. Crucially, this second olfactory organ is also fitted with chemoreceptors that can detect pheromones, those chemical signals essential for social and sexual communication within a species.

Once detected, these signals are processed in a specific part of the olfactory bulb and then pass along a neuronal pathway that leads directly to the hypothalamus, immediately stimulating certain behaviours as they hit the brain. When you see a dog with its lips curled up and nostrils flared, you are witnessing the flehmen response, the moment when a dog is opening up its mouth to draw in air to its vomeronasal organ. Similar behaviour can be seen in many other animals, including horses, deer and sheep.

This reflex increases the exposure of the vomeronasal cavity to odour molecules, opening up two tiny ducts in a dog's palate, behind its incisors, thereby heightening its ability to detect and identify an odour or pheromone – and react to it. Generally, the flehmen response is triggered in the presence of the smell of urine or scent from genital regions and is often accompanied by licking. It's how dogs suck up scent.

A dog's life

What does this olfactory acuity do to a dog's life? Apart from compelling them to mark territory on their daily walks, dogs depend on their sense of smell for many interactions and to determine social standing in the pack, even for fleeting acquaintances on their walks.

So, all that sniffing around another dog's butt does have a purpose. Their keen sense of smell draws them to rear ends, as that's where they can pick up all kinds of important and urgent messages. By sniffing this area, a dog can determine gender, health, diet and also another dog's dominant or submissive status. More importantly for the procreation of their genes, they can also detect reproductive status.

As anyone who's had a bitch on heat will know, male dogs are particularly receptive to an estrous female. Once a male dog picks up this scent, he will follow it over long distances. He will also do his utmost to cover the trail to prevent other dogs from competing with him, by urine-marking over her urine spots. Such scent behaviour will help him monopolize the female and thereby optimize the chances to transfer his genes to the next generation.

Most of the information that a dog may be interested in is held in the glandular secretions released by glands in a dog's anal sac, as well as in the sebaceous glands nearby. All at the rear. Much of this chemical information is detected by the vomeronasal organ and transferred to the brain.

These olfactory systems and pathways all contribute to a dog's sensitivity to odours and unique capability to detect a target odour among a multitude of other scents. Humans discovered this ability early on and learned how to make use of their sense of smell to our own advantage. But where and when did this special partnership start?

CHAPTER 3

What's with the wolves?

Looking at a husky, it's easy to believe that such dogs are descended from wolves. But a modern toy poodle? Or a chihuahua? In fact, DNA studies have determined that all domestic dogs share a common ancestor: they are all the tame descendants of the grey wolf.[6] Why they were domesticated is the subject of much speculation, and there doesn't seem to be a definitive answer. It is known that dogs were the first animals that humans tamed and might have provided both alarm and hunting services in exchange for food and security, but the details of the transformation from wild wolf to domestic dog remain elusive.

Domestication may have been an accident waiting to happen, building up over time. It's speculated that wolves were probably trailing humans for their scraps and the more docile ones got closer, were fed and survived, passing on submissive genes to their offspring, which resulted in the domestic dogs we know today. It's also unclear when or where exactly wolves teamed up to walk or work with humans. It may be around 20,000 years ago – or even 40,000. Selective breeding took care of the poodles and the labradoodles.

Whenever it was kick-started, this long-term unlikely partnership is rooted in a combination of our visual skills with the olfactory ones of the wolves.[7] As well as their ability to pick up on human social cues and an important hormone. They seem to be able to exploit our own bonding mechanism via the hormone oxytocin, the one that triggers the tight emotional bond between a mother and a baby and other relationships of trust. It's thought to be released when people, especially a mother and baby, gaze at each other. One study suggests that dogs have piggybacked on this mechanism to exploit it to create their own emotional bond and deeper attachment with humans.[8] That adorable look they adopt may have a higher purpose.

Hunting may have been one of the first areas where the wolf-human intimate bond took off, and it still remains a key part of our partnership with domesticated dogs today.

From anecdotes to research

From working with my own dogs, I have anecdotal evidence of their uncanny ability to track down both healthy and injured or dying animals. Once they pick up the track of a running deer or the scent of blood, they will go in one direction, double back, circle, sniff the air, sniff the ground, and lick surfaces to hone in on the source. My dogs' focus in circling in on the source is backed up by research, albeit not entirely consistently.

One study on domestic dogs, for instance, suggests that dogs are not always reliably able to follow a trail in the right direction.[9] Their ability seems to be linked to age, nature, gender and breed. Which dog is best for tracking?

The name is probably a giveaway, as is the number of scent receptors that it has: it's the bloodhound we mentioned above, with its 300 million receptors.

Other trials have determined that dogs can in fact follow a trail, especially if they have been professionally trained to do so. They may also however be picking up signals from the trainers to help them determine in which direction to head. Nevertheless, dogs can detect and follow an air scent, picking up odour molecules in the air, a ground scent, picking up footprints and the by-products of the steps in the disturbed ground, or a combination of the two: track scent. By sniffing around the source, they can determine in which direction the scent is fading away, and therefore not of any interest, and where it is at its strongest, and therefore will take them closer to the source.[10]

Remember those independent nostrils? They help a dog to sniff in different directions at the same time, so they can keep track on different strengths of an odour and pick out a fresh trail from an old one. Trailing dogs can pick up a scent from a distance, possibly even from over a mile away (especially when there is a female in heat involved).[11]

Impressive in action

Working dogs have become an intrinsic component of many walks of life, including law enforcement, military manoeuvres, search-and-rescue operations, as well as for medical and biomedical applications and emotional support services. They can even sniff out agricultural diseases.

Their ability to be trained to track down a target is astounding, and worthy of many short, feel-good news items. They tell of dogs that can follow trails over a week old, accurately pinpoint dead bodies in water or pick up miniscule traces of scent from earthquake or avalanche victims. They include stories of dogs that can detect minute quantities of explosives, as well as firearms, narcotics, and even computers.

It's no wonder scientists are trying to reverse-engineer the olfactory system in a dog to use for our own benefit. Or make more use of dogs to improve our own chances of survival.

Currently, there are numerous biomedical trials underway, looking, for instance, into how dogs can scent cancerous tumours in humans by sniffing out and identifying biomarkers associated with different types of cancers – before they would be detectable by other means. Dogs are already helping patients with diabetes or anxiety issues to get on top of their conditions. We delve deeper into this fascinating topic in Chapter 14.

Chapter 4

Birds Can't Smell, Can They?

There was a time, not too long ago, when a chapter on birds in a book about olfaction would have been dismissed, even ridiculed, by many in the science world. Birds were believed to be anosmic – unable to smell. The general consensus was that birds depended on their visual and acoustic skills to survive.

Anyone who has ever been woken by a dawn chorus would never doubt the importance of birdsong to any bird's survival and reproduction rate. Nothing quite marks the beginning of springtime more clearly than birdsong. And anyone who has witnessed – even if only on YouTube – a peregrine falcon swooping in on its prey with sharp vision and pinpoint accuracy from a great height and speed would never doubt the importance of visual acuity in birds.

Bright plumage colours in many bird species seem an obvious confirmation of what must be their excellent vision. Complex songs and peculiar dance performances confirm the important combined role of sight and sound to their territorial defences and mating rituals. But many have long doubted that birds exploit smell for survival. It just didn't seem to fit into their multi-sensory communication system. Why not?

One man can take much of the blame for this blunder: John James Audubon, an artist and a renowned ornithologist who died in 1851. Back in the 1820s, he claimed to have sure-fire proof that the most widespread New World vulture species, turkey vultures (*Cathartes aura*) or "buzzard" in US English, had no sense of smell at all. After playing a kind of dead-hog hide-and-seek with these vultures in their favourite dining territory, he determined that if the carcasses were hidden from view, under thick brush, the vultures would miss them. They consistently failed to home in on the hidden delicacies. However, they did swoop down if the carcasses were laid out in the open. It seemed fair, therefore, to conclude that these vultures operated by sight alone. Or was it?

Even at the time, his claim was controversial. Until his findings were published, there had been a common understanding that vultures were scavengers who were attracted by the nasty smell of death and decomposing corpses. Almost immediately, other scientists questioned Audubon's conclusions and conducted their own experiments. One creative attempt involved placing a canvas painting of a dead sheep, blood and gore painted in bright red, in full view of vultures. Rather comically, the vultures became obsessed with the painting, consistently pecking at it until they grew tired. Even when offal was hidden nearby, the vultures still headed straight to the painting, seemingly guided by sight rather than smell.

These experiments, organized by John Bachman, a Lutheran clergyman and naturalist based in Charleston in the US, appeared to back Audubon's results.[1] Together, their conclusions convinced the scientific community that birds were anosmic and driven purely by visual cues for foraging, which ultimately led to a dearth of studies on bird olfaction. It took over a century for other ornithologists to seriously question Audubon's conclusions and start looking into bird olfaction again.

The game-changing discovery

One of the first to cast serious doubt on Audubon's experiments was the ornithologist Betsy Bang of the Johns Hopkins University in the US. Her pioneering work started in the 1960s and was a game-changer for the field. She measured the olfactory bulb in the brain of over 100 birds, including the turkey vulture. Audubon might have been embarrassed to learn that the birds had large well-developed olfactory brains, which, Bang concluded, must reflect the important role that olfaction played in their world. It would take another twenty years, however, until another expert produced hard evidence to show that turkey vultures do in fact sniff out their lunch.

The ornithologist David Houston[2] provided perhaps the earliest and best evidence that Audubon and the other scientists had been guilty of a terrible oversight. In his experiments in the 1980s on Barro Colorado Island, Panama, Houston placed dead chickens either hidden or in full view in areas where the turkey vultures were known to go scavenging. The foraging birds were able to find the hidden delights even in dense forest, but they were clearly drawn to the carcasses that had only begun to decompose. There was an ideal Goldilocks level of decay and decomposition that attracted them: not too fresh, and not too rotten, ideally just around one day old.

Turkey vultures do prefer their food deceased, but they want it fresh, Houston concluded. In Audubon's experiments, the hogs had smelt off. The vultures had picked up that the carcasses were decidedly passed their "best-before" date and chose to ignore them. What Audubon took as a sign of anosmia was just proof that the buzzards were fussy eaters, and that their keen sense of smell would lead them only to the tastiest of morsels.

In the meantime, thanks in part to more advanced equipment and expert dissections, researchers have established that turkey vultures are even more sensitive to smell than other vulture species. They have a sizeable nasal cavity and an olfactory bulb that is four times larger than the bulb in a *Coragyps atratus*, or black vulture, for instance. And theirs contains double the amount of mitral cells, the ones that are sending olfactory messages onwards in the brain, even though their brains are twenty per cent smaller than that of the black vulture.[3] Altogether, this ensures that the turkey vulture has a superior sense of smell and is better able to detect volatile odorants emitted from carcasses. It still doesn't explain why they would choose to attack a painting of a dead sheep.

A vulture of the seas?

Out at sea, another bird is hard at work foraging for food. The albatross is built for the ocean, inspiring awe rather than disgust. With an average wingspan of just over three metres, their wings enable them to soar and glide across incredible distances over our oceans, spending years without ever touching down on land. Their wings are not the only characteristic that sets them apart from other birds. Much like the land-based turkey vulture, the albatross also has a well-developed olfactory system.

The albatross is what is known as a tube-nosed bird, or a *Procellariiform* (petrels and shearwaters are two others). Bang ranks these ultimate long-distance birds among the top 12 for olfactory prowess. Its nostrils stick out slightly from both sides of the top end of its bill and are among a number of reasons for its heightened sense of smell. A large olfactory bulb is another. How do we really know that the albatross that follows the fishermen at sea is following its nose and not just the boat?

We have one influential scientist in particular to thank for this knowledge. Gabrielle Nevitt, a US expert in bird olfaction and sensory ecology. In the 1990s, Nevitt led a number of olfactory studies on the albatross at sea. An unlucky personal injury on board one expedition led to a lucky encounter with another scientist on another, and eventually to her trailblazing experiments at sea.[4] The scientist she bumped into was investigating dimethyl sulfide (DMS), a gas emitted by phytoplankton, the microscopic plants that live just at the ocean's surface.

Nevitt knew that krill ate phytoplankton, and that, as they dined, DMS would be released into the atmosphere. She also knew that the albatross was fond of krill. Putting two and two together, Nevitt decided to test whether it was this gas that attracted the oceanic bird to its food. Once she had the opportunity to test her hypothesis, she was able to show that an albatross would double back during flight if DMS was released behind it over the ocean, but not if other control substances were. The gas clearly helps the albatrosses to find food.

If this seabird's acute sense of smell helps them find food, could it also be the reason why migratory birds can find their way home? It's hard to imagine that an odour plume would provide reliable directional information at sea – winds pick up, change course, the turbulent sea churns up and storms take hold. But how else would a seabird navigate over a featureless and monotone ocean? Out at sea, landmark cues are obviously missing.

Do these birds actually have a different map to the one we generally have in mind? Could the birds imprint an olfactory airborne map of where they have been and where they need to go as they fly away from their nest, much like salmon do in water (see Chapter 5)? What other kind of navigational toolkit would work in such conditions? This is a mystery that scientists still haven't yet completely solved.

Navigational tools

Could the Earth's magnetic field provide the answer? While some birds do use magnetic cues to decide which direction to take, studies have shown that *Procellariiformes* don't rely solely on this information, if at all. When the magnetic sense system is disrupted (for instance, by fixing a magnetic device to the top of a bird's head), an albatross can still find its way home, even in the absence of visual cues, landmarks or celestial cues.[5] The magnetic field can't therefore be the full answer.

This leads scientists to turn back to the theory that olfaction plays a bigger role in navigation and ensures that the albatross can home. In studies with shearwaters, who, like the albatross, are *Procellariiformes,* it was determined that displaced birds could best find their way home if their olfactory senses were intact.[6] It's possible that they remember smells as a kind of directional aid, and then head in the direction of the smell they remember. Not that these birds are following just one smell, they must be able to tell where they are and where they need to go by the smells and the different intensity of the odour plumes as they fly.

Many early experiments involved sensory deprivation – often temporary – of some kind. Other less invasive research using a mathematical analysis of the paths flown by shearwaters suggests that the birds do rely on olfactory-cued navigation to find their way back to their nesting sites.[7]

To make up for what it lacks in landmarks, perhaps these birds can create an olfactory landscape of the sea as they zigzag across the ocean, enabling them to follow erratic odour plumes of the next dinner. Or back to their nesting area. Other research carried out in nesting sites appears to show that the albatross depends on multimodal sensory mechanisms – visual and olfactory – for survival.[8] While there is plenty of evidence of the use of smell to find food, there does not, as yet, appear to be conclusive proof that migratory birds can depend on their sense of smell alone to find their way home.

But perhaps there is some convincing research on pigeons.

The original home birds

Someone who hates to leave the house is known as a "home bird". According to the Collins online dictionary of English, the term "home bird" was influenced by the homing pigeon's desire to stay close to home, and also by its remarkable natural ability to find its way back home across long distances. Homing pigeons home, but they don't migrate. Migratory birds, in contrast, only survive if they travel between their winter homes and preferred nesting sites during the breeding season. It's us humans who force pigeons to travel long distances for our own benefit.

In some cases, we capitalize on their weird homing instinct purely for our own entertainment. In the UK, and, to a lesser extent in Germany too, pigeon racing is still a serious sport, although by far not as popular as it was amongst the working classes after the Second World War. The magazines *Racing Pigeon*, first published in the UK in 1898, and *Die Brieftaube,* first published in 1883, are both still in print today. Apparently, Queen Elizabeth II has long been a keen pigeon fancier.

Before we look at the research, it's worth considering the diverse role of the pigeon in our society. In most cities, pigeons are regarded as a nuisance or vermin, much like a flying rat. But they never seem to fly very far voluntarily, even in the most dire of circumstances.

During The Great Fire of London in 1666, for instance, Samuel Pepys observed that "the poor pigeons... were loth to leave their houses", while all around them humans and animals took flight.[9] One single pigeon causes an existential crisis in Jonathan Noel just because it hangs around outside his home one day. (Admittedly, Noel is only a character in *The Pigeon*, a novella by the German author Patrick Süskind, but it's a fascinating take on such a bird's influence.)

These creatures are intelligent. For a start, they are one of the few to pass the mirror self-recognition test (along with the apes and elephants). Combine this intelligence with their instinct to home and you have some very useful creatures.

Pigeons have been relaying messages between different parties since at least the Roman Empire. Reuters News Agency kicked off their services using pigeons to fly stock prices between Aachen and Brussels in the 1850s.[10] During the First and Second World War, these birds were used to send messages behind enemy lines – or to report back on a pilot's whereabouts if a plane were shot down on enemy territory. The surviving pilot could attach details of his position on the pigeon and release it, sure in the knowledge that it had a good chance to reach its loft in the military base back home. Fascinating details from the Second World War are revealed by Gordon Corera in *Operation Columba – The Secret Pigeon Service.*[11]

The military knew that homing pigeons – *Columba livia* or the rock dove – were able to expertly navigate the skies, even if they didn't understand how they did it. The task of finding that out was left to the scientists. What belongs in this bird's range of tools? Olfaction, obviously.

Much of what we now know about the olfactory skills for navigational purposes of these pigeons really kicked off in Italy. Floriano Papi, a scientist based at the Scuola

Normale Superiore in Pisa, was fascinated by animal navigation, and, most of all, by the navigational skills of homing pigeons. He was the person who came up with the "olfactory navigation hypothesis", a belief that these pigeons relied on odours to find their way home.

Until Papi's hypothesis, other scientists had concentrated on studying this bird's use of the geomagnetic field, the stellar sky or the sun to determine compass direction.[12] Avian olfactory capabilities were rarely reviewed, again because of Audubon and his convincing reports.

Papi performed a number of intricate experiments with pigeons. Several involved the severing of their olfactory nerves in order to destroy the birds' keen sense of smell. The pigeons who were deprived of their olfactory senses were either unable to find their way home at all or took longer than the pigeons whose nerves were still intact. Papi didn't leave it there; he continued with a number of ingenious experiments that tested the pigeons' sense of smell and orientation abilities to the limits.

His research seemed to consistently show that the pigeons depended to a large extent on their sense of smell to home. He believed that pigeons would build an olfactory map at their loft, by associating the environmental odours carried by the winds with the directions from which they were blowing. This meant that the pigeons could make use of these maps to find their way home when they were released. Despite the size and breadth of Papi's experiments, some scientists still doubted that smell was the dominant force for pigeons to navigate the skies.[13] Maybe the birds were just traumatized from all that invasive prodding and surgical procedures.

Modern technology and less intrusive techniques have since given scientists more possibilities to put the pigeon's senses to the test even further and more reliably. Using GPS/GMS trackers, scientists can now follow the flight paths of birds more closely, and the use of zinc sulphate treatments can render the birds anosmic temporarily – and therefore more humanely.

In one study using such technology and techniques, a comparison was carried out between three sets of pigeons. The first group had all their senses intact throughout the whole experiment, a second group were rendered temporarily anosmic at the release site just before being set free, having been able to smell the air along the journey out, and a third group were rendered temporarily anosmic before they were transported to the same site as the other birds and set free.

Much like the studies with the shearwaters, this one determined that the third group, who had been made anosmic before being transported, performed the worst in the homing flight. They did find their way back, but they appeared more disoriented as they set off, heading in the wrong direction at first. They also needed more breaks along the return flight, possibly to get their bearings. They also took a longer, roundabout journey home. This tracking study provides very clear evidence that a pigeon's ability to find the way home is enhanced by its ability to detect environmental odours both at the home loft and along the journey.[14] I'm not saying it's their only tool, but it's definitely one that they use.

Birds make use of smell to find food and their way home. What about for mating?

Mating season

Choosing the right mate determines the reproductive success of an individual. Sometimes, monogamy has evolved in avian species, particularly in areas where the parents have to travel great distances (by great I mean thousands of kilometres for weeks and months at a time) to forage for food while the other partner stays home with the young or brooding. Penguins are one such species. Monomorphic (both sexes look very similar in size and appearance), monogamous and living in huge colonies of thousands of birds that look alike – how do they choose a mate or even find their way back to the right partner and their young?

Penguins certainly use acoustic cues to find their mates in the colony, as visual cues aren't helpful in the masses. This we know from a number of works, notably from Thierry Aubin, a specialist in vocal communication processes of animals in noisy environments,[15] and a senior scientist at the French National Centre for Scientific Research (CNRS). Olfaction, however, seems to play a role, too.

One study involving Humboldt penguins at the Brookfield Zoo in Chicago showed that these creatures use the sniff test to identify their relatives, probably making use of this technique to avoid inbreeding and also to track down their nest mates in the colony. In the study, scientists recorded the behaviour of the penguins as they were exposed to scents taken from the preen glands of other birds in their colony. The zoo knew whether the birds were related or not and were therefore able to observe if this fact made a difference to the penguin's reaction to the smell.

What was interesting in this study was that penguins that had partners lingered longer where they sensed the smell of their mate or a familiar scent, but not where the odour was unfamiliar. In contrast, penguins without a partner hung around longer where the scent was unfamiliar, which meant that it had not been taken from a close relative or their kin. This behaviour seemed to suggest that they were interested in the scent that might lead to a suitable mate, and not to a close cousin.[16]

Kin recognition

It's not only penguins that seem to exploit olfactory cues to avoid inbreeding and identify kin. Studies have shown that zebra finches, small songbirds known to communicate through learned vocalizations, seem to be able to recognize their relatives by smell. Barbara A. Caspers, Professor of

Behavioural Ecology at Bielefeld University, is an expert in this field and has researched this area in depth. Results of one study suggest that zebra finches can identify their kin by smell alone.[17]

The experiments involved placing finch fledglings into different broods as soon as they hatched, and then, at around 20–23 days, presenting them with a task – choosing between two odour samples. One was from the biological parents, the other from the "foster family". The zebra finches invariably chose nature over nurture. The team of researchers believe this indicates that the songbirds exploit such chemical cues to avoid inbreeding – and to recognize their kin.

A more recent study suggests that parental odour is imprinted on zebra finches during the early stages of embryonic development.[18] Even when the eggs were fostered, which meant the fledglings had no contact with their biological parents at the moment of hatching, the zebra finches were still able to identify – and still preferred – the parental odours (particularly of the mother) over the foster family. The scientists argue that the telltale odours were absorbed during embryonic development, as the porous shells allowed the parental scent to pass through and imprint on the finches. The maternal odour appeared to have a stronger effect.

There is another possibility, one that the scientists involved in this study also consider but did not fully investigate. Preen gland secretions could also play a role in determining the smells that are absorbed in the natal nest and that could be linked to kin recognition.

Preening and mating rituals

It's well documented that birds preen their feathers to keep them clean so they can fly better. The preen gland, or uropygial gland as it's more formally known, is located close to the bird's tail and is the source of preen oil that birds need to clean and waterproof their feathers. It's their own personalized, private and seemingly unlimited supply of hair oil, so to speak.

Scientists now know that this preening oil has a specific blend of odours in many bird species. And that this scent could be affected by the bird's microbiome – the fungi and bacteria that live on a bird.

Research using dark-eyed juncos, a common North American songbird, suggests that the microbes that live on the birds' preen glands may play an important role in making the personal scent molecules of each bird. Scientists believe that the birds' microbiomes may influence both the odour and the behaviour it triggers in other birds.[19]

In the study, samples of bacteria on the preen glands were first taken, the glands were then injected with an antibiotic. When before and after comparisons of the preen oil were made, it was clear that the odour in the oil had changed after the treatment with antibiotics, and that the change had come from the missing bacteria. This missing component seemed to influence the songbird's reproductive success. The juncos are known to have evolved to mate only with birds that live in the same area. The songbirds that live in urban areas no longer mate with the same species that live in woodlands. It's possible, the scientists say, that the choice is made based on scent, as the city birds don't seem to like the smell of their country cousins.

In some cases, it's the sweet smell of tangerines that will attract the right mate. The crested auklet (*Aethia cristatella*) looks like a happy little punky penguin with a bright orange beak and a black tuff of feathers on the top of its tiny head, but it belongs to the *Alcidae* species. They live on remote, rocky islands of the North Pacific, and, like penguins, they live in massive, noisy, smelly colonies and travel huge distances across the ocean to feed. But they differ from penguins in one unique way. They give off a citrus-like smell of tangerines during the breeding season. Why?

Scientists have been able to pinpoint the exact area on the auklets where this citrus-scent emanates, from a patch of special wick feathers on the back or nape of their necks. An area that the birds will snuggle up against during mating rituals, an action which scientists have named a "ruff-sniff".[20] This seems to release the scent, which is dominated by aldehydes, and, along with it, an added bonus.

The odour may have a built-in insect-repellent function, which is important on a territory that is teeming with ticks and parasites. The mates that smell the strongest might be the healthiest, and therefore attractive to potential partners. Such smelly mates are probably tick-free and can provide their partner with added protection from the insects, by sharing the compound during the mating rituals, and by not bringing parasites into the nesting area. Experiments using smelly decoy birds using a synthetic mixture of aldehydes have shown that the auklets are attracted to the smelliest fakes, serving again to debunk that long-held belief that birds couldn't smell at all.

It's in the genes

While many scientists have taken a somewhat surface view of scents in birds, Silke Steiger at the Max Planck Institute for Ornithology (MPIO) has adopted a decidedly different approach. One area of her research has focused on the olfactory receptor (OR) genes in birds.[21] The number of such genes in a genome very likely indicates how many different scents an animal can detect or distinguish. Or at least provides a very good rough estimate.

When the researchers at the MPIO compared the OR genes of nine bird species, their studies found that the majority of olfactory-receptor genes were probably functional, but that there were considerable differences in OR gene number between the species. Interestingly, the highest number of OR genes correlated with the relative size of the olfactory bulb in the brain. The kiwi, for instance, which has the second largest olfactory bulb in birds relative to its size, was found to have a high number of OR genes. This is not surprising given the lifestyle and territory of the bird. Scientists believe that its ecological setting may have shaped its OR gene repertoire size, in a similar way to how it evolved in mammals.

This nocturnal ground-dwelling bird has nostrils at the tip of its long bill, which proves essential when foraging for food on the ground under the stars. As it has a very small visual field, one of the smallest ever recorded in birds, it seems logical that its sense of smell should have evolved to compensate for this deficiency. When kiwis probe the forest floor at night for earthworms, insect larvae and other invertebrates that rise to the surface at night, their noisy sniffing and snuffling is a sign that they are guided by smell, rather than sight. Their bill is also thought to be sensitive to touch. They like to wave it around in the air, like a wand, inhaling noisily in the process.

Birds and us

Kiwis, auklets, zebra finches, juncos, penguins and pigeons – they all are living proof that can refute Audubon's claims that birds can't smell. Why does this matter to humans?

Let's take the vultures. You may think that they are a bit of a sideshow for humans, but they have an important, central ecological role on our planet. These scavengers and their (to humans) disgusting habits help prevent the spread of bugs and pathogens among other animals, which ultimately helps reduce the likelihood of humans catching nasty diseases.

Pathogen transmission ends with these obligate scavenger vultures – that dine on carrion rather than hunt down live prey. When the vultures go missing from their usual territory, bugs and diseases seem to circulate more freely. In India, for instance, a massive decline in the numbers of the Gyps species of vultures in the 1990s resulted in an increase in rabies transmission in humans. How are the two connected?

First of all, massive is not an exaggeration – the vulture population went down by 90 per cent. Scientists discovered that an increase in the use of diclofenac, a non-steroidal anti-inflammatory drug, in livestock was fatal for vultures in the Indian subcontinent.[22] Vultures that fed on dead cattle that had been treated with the drug would die. Their kidneys would malfunction. As the vultures were dying out, more and more carcasses were left to rot, contaminating drinking water. Feral dogs took over the scavenging, increasing their numbers, but the dogs are not as efficient in disposing of the carcasses. Vultures would pick them clean, but the dogs left the yucky bits and pieces behind, and with them lots of bugs, which was not good for the environment. The dogs also picked up and passed on rabies and other diseases to humans.

Based on the conclusive finding that diclofenac was destroying the vulture population, the Indian government acted. An alternative drug was found that did not affect the vultures, and diclofenac was banned across the country for use in livestock. Ironically, the drug had been used with the best of intentions in many cases. Cows are considered a sacred symbol of life for Hindus, and as such their slaughter is taboo. To make the cattle's final days less painful, the drug had been administered to ease their pain. The vulture population has started to recover, the crisis, however, does not appear to be over yet. Similar drugs are still in use.

Upsetting the balance

This story is a stark illustration of just how intertwined decisions and actions taken in one section of society can end up upsetting the balance in another part of nature. But that also there are solutions that can help rebalance the losses. Vultures may not look pretty and may have some terrible table manners, but at least they help keep the ecosystem in a kind of balance. And they need their acute sense of smell to do so.

Another connection between human actions and birds and their acute sense of smell has yet to even hint at a positive outcome. Those beautiful albatrosses and their olfactory sensitivity to DMS are disadvantaged in a world that is becoming increasingly polluted with plastic.

Apparently, plastic easily absorbs and traps the smell of DMS if it remains in the ocean long enough – which may only be a matter of a few months. This scented plastic then entices foraging seabirds to eat this deadly meal instead of krill.[23] In an ideal world, this finding could feed into research on developing materials that don't pick up the scent in case they end up in the sea. Even better would be less trash in the oceans in the first place, as I argued in Chapter 1.

Seabirds and the vultures both deserve our respect. Their keen sense of smell has handed them an evolutionary advantage in their own ecosystems. Our meddling can all too easily tip that gift into a disadvantage.

Chapter 5

Who Smells a Fish?

Some of the most bizarre creatures on this planet are found in our oceans, rivers and waterways. Underwater is where you'll find self-sacrificing parasitic partners and vampire-like predatory suckers, as well as docile and not-so-docile masters of homing navigation.

Although each species has evolved very differently, they all exploit a keen sense of smell for survival in their specific environment.

Humans are only able to detect odours because volatile scent molecules have floated into our nostrils. There, they dissolve into the thin wet lining, the snot, coating our olfactory receptors, which are waiting to trigger neural signals. If a moist atmosphere is what helps us identify an odour a wet environment should be favourable to smelling. Try sniffing underwater, however, and you'll soon be spluttering for air before you can enjoy the scents of the sea. As we sniff, we also inhale into our lungs to breathe. We cannot separate these two functions. Your lungs would fill with water and you would die quickly.

Amphibians can clearly separate the two. As we evolved to live on land, our noses were modified to pick up smells only in the air. Frogs, toads and other amphibians, in contrast to humans, are fitted with two-chamber noses. This means they can still sniff in the air with one chamber, shutting it off when underwater, where they use the other chamber to detect smells without drowning. The air chamber has odour receptors sensitive to volatile odours in the air, while the aquatic chamber is fitted with olfactory sensory cells that operate only with water-borne molecules.

Not all mammals risk drowning when they dive and smell, however. One semi-aquatic mammal, the practically blind star-nosed mole, is able to sniff underwater without inspiring water. This creature's snout is about as weird as it gets. With its 22 pink tentacles, it does have a star-like quality about it. Sensitive to touch on land, underwater this snout becomes an olfactory tool. The mole exhales tiny bubbles out of its nostrils, only to re-inhale the same bubbles at lightning speed, along with any odour molecules they've captured along the way. Experiments using high-speed video footage show that the mole uses this technique to sniff out food.[1] Similar techniques are thought to be used by water shrews and some otters, but more research is needed to be certain.

These moles only stay underwater for seconds at a time. What about the creatures that spend much more time underwater? How do they master sniffing and underwater survival? How do they smell?

Fish-smelling anatomy

As any good fisherman knows, fish can pick up scents underwater from a distance. But underwater creatures are incredibly diverse. They don't all have the same olfactory anatomy.

Many species have nares or nose pits lined with cilia – microscopic hair-like structures fitted with odorant receptors. The nerves connected to the cilia lead from their olfactory organ that acts as an alarm connected directly to the brain, sending electrical spasms with information that is crucial for their survival. Even fish larvae can pick up scents. Four-day-old zebrafish larvae have motile cilia, which is the kind that can move. But the movement doesn't seem to be random. These cilia pulsate to a rhythmic beat, becoming microscopic turbines in the process.[2] As they work, they increase the flow and exchange of odorant molecules over the olfactory epithelium and then out to the sides. Such ciliary turbines also improve the sensitivity of the sense of smell in fish and their ability to detect and process odours. These turbines, it seems, are particularly useful in stagnant water.[3]

So, how do fish use these sensitive olfactory systems to survive?

Pheromone communication in fish

Recent research on the fish olfactory system has determined that fish have developed three parallel but distinct pathways from their olfactory epithelium. Each conveys specific information that triggers certain survival behaviour in response to odours. One is concerned with social cues (including those that warn of predators), the second with sex pheromones, and the third with food odours.[4]

Goldfish, one of the most studied of fish species, release hormones and metabolites to trigger certain behaviours in members of their species. Five hormonal products have been identified that appear to have a specific function. Electrophysiological recording from the olfactory epithelia of well over a hundred species of fish show that most detect hormonal products, although it's still not entirely clear how fish make use of the odorants. It appears that they do regulate reproductive behaviours. When male goldfish pick up the scent of postovulatory female goldfish pheromones, they will automatically increase their milt volume (sperm release). Interestingly, they'll have the same response to certain chemical cues from male competitors in their vicinity.[5] For this species, it seems survival of the fittest is decided by intense sperm competition. How does it look deeper down?

Self-sacrificing parasitic partners

A deep dive into our oceans uncovers a world that may seem at first to be the dead-zone of our planet but is in fact anything but.

Around or below a thousand metres underwater, in the mesopelagic zone, or twilight zone, and the bathypelagic zone, or midnight zone, there are hundreds of incredibly varied species of deep-sea creatures, among them the anglerfish. The humpback anglerfish is one scary deep ocean swimmer. Almost ghoulish in appearance, these fish look like the bad guys of the seabed. The females look the most terrifying. Usually much larger than the males, many are fitted with types of "fishing rods" on the tops of their heads that seem to glow in the dark.

Such bioluminescent devices mimic the movement of bright live bait in the sunless depths of the ocean, possibly luring their prey into the needle-like rows of teeth of this aptly named fish.

Once enticed into a mouth that can often open up wide enough to trap prey much larger than the predator itself, the female's teeth become the bars in her death-row jail. And its elasticated stomach expands to take in whatever prey comes her way. At the depths of the ocean there's no telling when a tasty morsel might pass by. When it does, the female anglerfish has to be able to attract, lure and trap it tightly – whatever the size.

The female deep-sea anglerfish has clearly evolved to survive in such an inhospitable environment. What about the males? This species is extremely sexually dimorphic, particularly when it comes to their signalling and sensory systems. While the females are larger on the outside, it's the males who, relative to their size, have the larger olfactory organs on the inside. This would indicate that they can pick up a trail of pheromones to track the female.

Studies suggest that the females drift along with the current, releasing pheromones as they cruise along. Ideally, they remain as stationary as is possible in the depths of the ocean, in order to increase the odds of them being spotted. The males, meanwhile, swim across the current, vertically and horizontally in a random pattern, until they pick up the pheromone field of the female. Once they pick up her scent, they will swim horizontally, following her trail. If he detects a weakening of her scent, he will zigzag in random directions until he picks up her pheromone patch again.[6] Much like a search-and-rescue activity, but this time it's search and reproduce. Visual cues are only useful up close at such depths, once the long-range chemical cues have served their purpose.

Finding a mate in the deep dark depths of the oceans can't be easy. Which is why once it has sniffed one out, the anglerfish dwarfed male bites down hard and may never let go. Or at least he won't until spawning can take place. It can turn into a parasitic union for life.

The male deep-sea anglerfish attaches to the female with his teeth until the couple gradually fuse together – into a single circulatory system, connecting skin and bloodstream. The longer he waits for spawning, the tighter the fusion.[7] In the process, he loses his eyes and all of his internal organs, including his olfactory organs will degenerate,[8] except the testes. Those are still needed, as his sole function is to produce sperm for his host. Think of him as a portable sperm bank. He's one determined sexual parasite, but he's often not alone. The promiscuous female of this species has been seen to carry up to six males at a time.

But don't feel sorry for the guy. If the male doesn't find a mate, he will probably die anyway. As already mentioned, there's really not a lot of prey at that depth. And while his olfactory system is exemplary, his digestive tract leaves much to be desired. It's so stunted that he can only survive as a parasite hanging onto a female partner. What a couple. What a sucker.

Vampire-like predatory suckers

Another underwater bloodsucker is the sea lamprey (*Petromyzon marinus*). As with goldfish, lamprey species rely on pheromone communication for survival. As with the male anglerfish, they have parasitic habits. At first glance, you might mistake a lamprey for an eel – until you notice its mouth. And remember that lampreys are the world's oldest living vertebrates (even if they have cartilage rather than bone), and that an estimated 300 million years of evolution probably separates the two.

Neurobiologically, the sea lamprey represents a proto-vertebrate. This creature is so primitive that it doesn't have a jaw and therefore cannot bite. Its mouth is permanently fixed in an open position. Resembling a suction cup, it's lined with rows upon rows of razor-sharp teeth and packs a toothy chisel-like tongue inside. It's ideal for chipping a hole in its host. Lampreys are suctorial creatures. They latch on to unsuspecting victims by gripping, and then they suck out fluids and blood to survive – usually killing the hosts in the process. Since their gill openings lead directly to their throat they can hang on and continue to respire as they suck.

It's estimated that each lamprey can kill up to 18 kilograms of fish, their preferred host, each year.[9]

Sea lampreys are monorhinic, which means they move scented water into and out of the olfactory capsule through a single nostril with each respiratory cycle.[10] Their respiratory flow is controlled by their velum, a muscular structure that contracts to ensure the flow of water through their pharynx for both feeding and respiration. Unlike the anglerfish males, who detect and track the scent of the female of the species, it's the male of the sea lampreys who secrete a scent – and the females are the ones who track the males down.

Much like salmon (more on them next), sea lampreys swim upstream to spawn. Behavioural tests have confirmed that spermiating sea lamprey males release a bile acid that acts as a potent sex pheromone, signalling its reproductive status and location to ovulated females. 11 This sex pheromone can act over long distances, in some cases even two kilometres away, luring the females into the desired location.

Such discoveries have led to experiments with pheromone trapping in areas where the lampreys are seen as undesirables. The traps are designed to work by emitting a biodegradable synthetic version of the appropriate sex pheromone in an area that is unsuitable for spawning. Experiments are still in their early stages, but the idea is that the females will be attracted by the decoy scent, only to end up lost and alone, unable to lay the thousands of eggs that they normally would. In areas where this invasive species is devastating native species populations, fishing communities and ecosystems, including in Lake Ontario, Canada, such an eco-friendly solution would be welcome. More on this in Chapter 14.

Docile masters of homing navigation

Unlike lampreys, which take any stream that has a suitable breeding space, salmon always seek out the "birth" freshwater stream to spawn. This is the place where they may have spent only a few days or a few years, depending on the species. It's thought that they use a combination of visual stimuli, electromagnetic cues and their acute sense of smell to find their way back. Their homing ability is extraordinary.

Salmon start off in freshwater streams, where they may live for a few days, or a few years. Whatever the duration from hatching to smoltification (the process of physiological change they need to undergo to adapt to living in seawater), it's believed that the salmon use this time to imprint a chemical map of their home base on their brains, which they refer to when it's time to spawn. Which could be anything from two to eight years after the moment they first migrate to salty marine water, and, depending on the species, after covering hundreds or thousands of kilometres. It goes without saying that it's a huge challenge for them to return. How do they master it?

Scientists believe that salmon may use the earth's magnetic field for compass orientation back to their native stream. Geomagnetic cues may guide navigation. Salmon must be able to recognize visual cues. Perhaps they can track the passage of time. But to home in on the exact location of their home riverbed, salmon rely on their sense of smell. It's not known exactly how sensitive salmon are – maybe they can detect one odour molecule in concentrations as low as one part per million, or even in one part per trillion – but we do know that salmon can detect the exact odour molecules that match the river where they were spawned.

Experiments have shown that salmon imprint olfactory memory of natal stream odours as they migrate downstream. Their home base no doubt has a unique combination of aquatic plants, fauna and soil, which combine to create an equally unique scent. On their return, they can recall this stream-specific odour information to identify the route they need to take during upstream migration for spawning. Salmon really do have a chemical map of their home river.[12] How do they detect these chemical cues and decipher them?

Salmon have nasal cavities on both sides of their heads, below the eyes, that contain around a million tightly packed olfactory cells. Unlike lampreys, salmon have olfactory cells with cilia – those hair-like structures with odour receptors. They probe the water for the odour molecules that they have evolved to identify. Each odorant has a specific shape and composition that matches only one type of receptor. Once the pair matches up, much as a key fits into a lock, a chemical impulse is triggered, moving out of the nasal cavity and into the olfactory bulb of the brain, which processes the incoming information about the salmon's environment. Neurons inside the bulb itself process and organize this incoming data, sending it on to appropriate areas in the brain that has evolved to deal with these triggers.

Their sense of smell is not just useful for homing. Apparently, it comes in handy for detecting and avoiding predators, even in juvenile fish. Lab experiments have shown that salmon can recognize a threat by its smell. Young salmon will avoid the areas in water tanks that had been scented with diluted otter faeces. The fish avoided those areas only when the otter in question had dined on salmon beforehand. If the scent was from an otter that hadn't feasted on their species, the salmon didn't avoid that area. This suggests that it is the smell of the diet of salmon, and not the smell of the otter itself, that acts as an early-warning trigger to the fish.

It's not important what creature presents as a danger, what matters is the fact that it's eating the salmon's family members and relatives, no matter how distant they might be![13]

Interestingly, some creatures will go even further to avoid predators. They'll use chemical camouflage to survive. It's common knowledge that coral-dwelling filefish will disguise themselves visually, so they resemble coral and fade into the background. But their predator avoidance mode also includes chemical deception. Chemical crypsis – the ability to avoid detection by a predator – in filefish also involves emitting the scent of their coral diet. This acts as a mask from their own predators, who aren't in any way interested in eating coral.[14]

Not-so-docile masters of homing

So, prey can use smell to avoid predators. How do predators use smell to zoom in on their prey? How about those underwater creatures at the top of the food chain that have the worst reputation? Sharks.

In total, there are over 500 species of sharks. Most tend to be oceanic in nature. Many travel thousands of kilometres each year, much like some species of salmon, to return to their natal environment. Many migrate vast distances across the oceans, exploiting currents, including the Gulf Stream, to undertake their clockwise transatlantic journeys. Along the way, each species uses visual cues, electroreception (the detection of electric fields, currents or pulses) and smell to varying degrees, depending on the environment. They are survivors – in deep, dark oceans, in shallow seas, and even in river systems.

The tagging of sharks has provided fascinating insight into how far sharks of different species tend to travel – and into the senses needed to find their way back home. Monitoring projects off the coast of Florida, for instance, determined that blacktip sharks (C. *limbatus*) depend on a combination of senses, including smell. If their nares are blocked, they have difficulty finding their way home.[15] The absence of olfactory cues stopped some sharks from homing completely, and those that did manage to home took longer to travel the distance back compared to sharks who had been released at the same time with their olfactory senses intact. They also displayed changes in their normal behaviour once they arrived back. They were less likely to stay in the home area.

However, it's not clear which chemical cues these sharks would have exploited on their journey under normal conditions. Maybe they were unable to clearly identify the signature scent of their home area. It's possible that they have an imprint of this chemical map in their brain, much like salmon, but were unable to detect it because of the blocked nares. Or they may have been unable to pick up on the pheromones of their fellow sharks, which confused them. What this particular research showed is that while olfactory cues are important, they are not the only sensory signal involved in homing. Geomagnetic orientation may play a role, or tidal currents could provide important information and facilitate movement in a certain direction. Maybe the fact that the signature scent of their home region was missing from the current also confused them. It's hard to know for certain.

What's clear is that this inbuilt array of effective sensors makes sharks ideal foraging creatures for deep-water habitats. Which is where they often prefer to be – again, depending on the species. Often, their sensors are amazingly adapted to find prey in low-light levels.

Their ampullae of Lorenzini, the jelly-filled pores on the snout, are the key to their unique specialized system of electroreception. It senses even minute muscle contractions of prey making a move to flee. This system helps them detect prey even in an environment where they cannot see at all. It also picks up the vibrations of distressed or injured creatures.

Sound waves travel well underwater and register on the shark's lateral line system, a sensitive network of tubes filled with fluid that starts at the snout and extends along both sides of its body, just under the skin. Its pores allow the flow of water to enter these tubes, which are lined with hair-like structures that are sensitive to vibrations in the water. This system guides the shark towards the general vicinity of its prey, even in the darkest depths of the ocean.

Great white sharks often swim in clear waters and therefore have good eyesight. They'll probably see you before you see them (but not usually if you're more than ten metres away). Will they smell you, too?

The murkier the water, the more likely it is that their sense of smell will kick into action. No doubt the image you have in your mind right now is that of a tiny trickle of blood making its way to a shark's nostril, only to trigger a blood-curdling predatory shark attack from a great distance. That's a myth. There is no firm evidence that sharks actively hunt down people by their scent. A shark attack on a human is usually a case of mistaken identity.

Dramatic scenes of actual shark attacks on people are nevertheless hard to forget. Such as the time when Australian surfer Mick Fanning was attacked by a great white shark while competing at Jeffreys Bay in South Africa in 2015. He may have walked away unscathed after punching it hard, but that doesn't make the attack any less terrifying. But such attacks are rare. One reason for this could be quite simple.

Sharks eat a wide variety of food, such as plankton, fish, crabs, seals and whales, but they are picky eaters.

They prefer food that is high in fat, which is why seals are often on the menu. If they nibble on something that tastes unusual or unfamiliar – such as a human – they often won't continue with their feast. Admittedly, this offers little consolation, as one nip is usually enough to be fatal anyway. Even if most people are not swallowed, they'll probably bleed to death.

How do they track down the prey that they do want to eat? Sharks do have an incredibly sensitive olfactory system. As they swim, water is constantly flowing into their nasal passages, or nares, automatically bringing a constant stream of scents from their surroundings into the olfactory canals and nasal sacs. This stream flows over the olfactory lamellae. In sharks, these thin sensitive layers of folded tissue ensure that the scent lingers in the nasal sacs for longer, which means these creatures have a better chance to detect odours. Odours that pass over the olfactory lamellae stimulate sensory cells that then pass on messages to the brain.

There has been a lot of speculation about just how sensitive sharks are when it comes to smelling, particularly on the acuity of the white shark, but no research has yet confirmed for certain that they smell any better than other underwater creatures. As their olfactory system takes up around two-thirds of its brain, it's speculated that they must be more sensitive.

Calculations estimate that a shark can detect some scents at concentrations as low as one part per 25 million. This is probably the equivalent to about half a kilometre away in the open sea.[16]

Crucially, they are known to be skilled in directional smelling. They follow their noses. Directional smelling helps sharks track down the source of a smell with more precision. They are able to isolate incredibly quickly whether the scent is coming from the left or the right, and then head in that direction. It's the reason why sharks can appear to sway as they move through water. The movement helps them to pinpoint the precise location of their prey. Left or right?

Hammerhead sharks seem to benefit most from this feature. The anatomy of their flattened head, known as a "cephalofoil" in scientific circles, makes them instantly recognizable. It also gives them uniquely widely spaced nares and the edge over other sharks when it comes to tracking and zooming in on prey. As their nares are located at a larger distance apart, almost along a line on their wide snout rather than on either side of their heads, the time it takes for them to notice that the strength of an odour is stronger from a different direction is short. They have smell in stereo.

This, researchers say, means they perceive a bilateral time difference at a smaller angle or at a greater swimming speed than other sharks.[17] They can therefore steer and attack at higher swimming speeds. They appear without warning. Unless, that is, its potential targets secrete *Schreckstoff*. This German word is best translated as "repellent" or "fear substance". More colloquially, it's "scary stuff". In the olfactory world, it's the term used to describe a chemical warning signal, or alarm pheromone, in fish.

The fear and fright factor

Most fish are social creatures and swim about in shoals. There is safety in numbers, after all. This can work particularly well in their favour when *Schreckstoff* is part of the group's defence mechanism. Fish emit it when they are injured – or being eaten – and it seems to warn the rest of the group to scarper.

Karl von Frisch, an Austrian animal behaviour scientist, first noticed the effects of *Schreckstoff* after placing a fish into a tank of other fish. For the purpose of an experiment on hearing impairment, the fish had a severed sympathetic nerve and, obviously as a side effect of this procedure, an injured tail. Frisch noticed that the other fish immediately all started showing signs of stress when the injured fish joined them in the tank. Consequently, he changed the focus of his experiment and decided to investigate what was behind their reaction. He coined the term *Schreckstoff* to describe the trigger, a chemical compound that literally warns the group that there is danger nearby. That was in 1942.

In the meantime, many other researchers have investigated the concept further. It's believed that the release of *Schreckstoff* in fish could be involuntary. This means it's likely to be a passive response to a physical injury caused by a predatory attack (such as Frisch cutting the fish's tail). While the scent of injury would attract other predators, the same scent seems to warn its group of a danger.[18]

A research team in Singapore recently reported that they'd pinpointed the exact *Schreckstoff* compound in zebrafish: a sugar-like molecule named glycosaminoglycan chondroitin.[19] It exists in various forms across living creatures, including in cartilage. (You can also get it as a dietary supplement in tablet form if your own cartilages are playing up.) In a series of experiments involving a selection process using a variety of chemical compounds from ground-up zebrafish, the scientists were able to identify this glycosaminoglycan chondroitin as the one that triggered predator-avoidance behaviour in the zebrafish, which manifested by them either staying near the bottom of the tank, darting back and forth, or swimming slowly for a while only to shoot off in different directions repeatedly.

What's interesting is that the scientists have been able to demonstrate that only one part of the olfactory bulb in fish reacts to *Schreckstoff*. It doesn't react to other chemical cues. The question remains, however, as to why fish send out this alarm at all. Is it a last-ditch altruistic attempt to save their fellow swimmers? Or is it that the fellow swimmers are likely to be genetically related and therefore it's worthwhile warning them so that can pass on the family's genes at a later point in time? It might also be that all fish have evolved the ability to pick up and process this *Schreckstoff*, and are therefore the real survivors of their species.

Mammals in water

It's clear that fish have an excellent sense of smell and use it to smell in water. What about the mammals that, during the course of evolution, returned to the water environment? In contrast to fish, they still breathe air. We saw how the star-nosed mole has developed a bubble method to track scent trails to survive. How about seals and whales, who live their lives predominantly or totally in water? Many studies throughout history have claimed that these marine mammals have no or, at most, maybe a rudimentary sense of smell. This impression still holds true for the toothed whales, such as dolphins, orcas and sperm whales. During evolution, they seem to have lost the neural hardware to smell.

Baleen whales, however, are a different story. Recently, they were shown to possess both the neural and biochemical structures to smell.[20] Scientists who joined Inupiat hunters during their annual subsistence hunt were able to both dissect the brain of the whales and take samples to investigate if they carry genes for olfactory receptors. In the laboratory, the scientists discovered that the brain had clear neural connections to the nose and that genes for receptors were indeed there. Similar observations have been made in other baleen whales. So why would they need to smell? Well, this could connect back to the albatross and other seabirds. They can all locate the smell of hydrogen sulphide, which is the telltale odour revealing the presence of plankton, along with what happens to be the baleen whales' favourite food, krill.[21]

A similar capability was recently demonstrated in seals, animals that have long been considered more or less anosmic – completely lacking a sense of smell. Experiments with harbour seals revealed that they can smell… dimethyl sulphide! Again, the important odour for sea animals. But this is not the only odour that seems to interest seals.

We've all seen those cute photos of female sea lions "kissing" their young. What we interpret as affection is in reality an important process of mutual identification – by smell. During the eighteen months that Australian sea lions (*Neophoca cinerea*) look after their young, nursing mothers often have to leave their offspring behind to hunt. When they return from their foraging expeditions, sometimes after days, the pups are unlikely to have stayed where they left them. Typically, they will roam around the colony, exploring the coves, often ending up in the midst of a group of other pups. How do the mothers recognize which one to suckle or feed?

Sight and sound do play a role. The female sea lions will use visual and audio cues to recognize their pups by their calls and appearance, but it seems that olfactory cues will provide the final close-range recognition of their pups, hence the appearance of "kissing" when they reunite.[22]

What exactly are they recognizing? Scientists studying the Antarctic fur seal (*Arctocephalus gazella*) discovered that the scent of both mothers and their pups on the sub Antarctic Island of South Georgia had similar characteristics, suggesting that the smell could be genetically encoded. This is significant for the survival of the species, helping prevent inbreeding and promote genetic diversity, the researchers believe.[23]

Even if marine mammals don't smell in the water, many of them seem to have retained the capability to pick up airborne cues, allowing them to pinpoint resource rich areas in the vast ocean. Why the toothed whales have followed a different evolutionary trajectory and seemingly lost their sense of smell remains, however, a mystery.

Scientists are still working hard to uncover this and other mysteries of underwater and sea creatures. What we do know is that different species use their sense of smell to home in on the right destination, hunt down prey, warn of danger, and seek out a mate or their offspring. As we humans try out scents to control or inhibit certain behaviours in these fascinating sea creatures, our own behaviour may also be threatening the very survival of many of them. The same holds for many other human activities as we saw in Chapter 2.

Chapter 6

For a Mouse, Smelling Is Everything

In *The Tale of Two Bad Mice*, Beatrix Potter lets Hunca Munca and Tom Thumb (the two bad mice) get extremely upset when they realize that the food items in the doll's house they invade are just made of painted plaster. Potter was probably not a real mouse biologist. If she had been, she would have known that looks bear very little significance to tiny mice. It's the smell that counts. The smell indicates what's edible (most things...), but also who's an eligible and strong partner, where your mother's nipples are to be found and who's a friend and who's a foe.

The mouse has followed humans to more or less every place on Earth and has been a constant nuisance ever since we started storing food for future consumption. Mice just like to eat the same things as we do. Originally, the house mouse stems from central Asia and arrived into the eastern Mediterranean around 13,000 BC. It then took until about 1,000 BC before it had colonized the whole of Europe.

Within about ten weeks of being born, the house mouse can start giving birth. It therefore has a very short generation time. And once it gets started, it can multiply rapidly. A female can have five to ten litters per year, each with six to eight pups. This means that six mice can become 60 in three months' time. You can see why mice easily become unwanted and numerous guests in our homes, especially when winter approaches and food sources outdoors are getting scarce. However, this very fast generation time has an upside. It's one of the factors that have made the mouse the model system of choice for many scientists working in biology and medicine.

When it comes to understanding the sense of olfaction two model systems have provided the most information: the fly and the mouse. Both systems have their pros and cons and complement each other quite well. The mouse has the advantage that, being a mammal, it's quite close to us. In this chapter, I'll look into the smell-dependent behaviour of the mouse and to some extent at the underlying machinery. Scientists all over the world have invested vast amounts of time, energy and resources to understand these systems, so I can only scratch the surface and transport us all into the vital aspects of the smelly life of a mouse.

The main nose of four

From a human perspective, the nose is the nose, and that's it. For most other animals, both with and without a spine, olfaction happens at several different places, and the mouse is a very good example of this. It has four different organs to smell with, each with its own unique morphology and functionality.[1]

Just as with us humans, the main olfactory organ in the mouse is located above the nostrils inside the skull, the nose. It consists of a highly convoluted mucosa that's proportionately many times bigger in the mouse. In the mucosa about 10 million olfactory smell neurons respond to different types of volatile molecules. The specificity of a certain neuron is determined by the olfactory receptor that it expresses.

The mouse has about 1,200 different types of these proteins,[2] which is about three times as many as we have. Each olfactory receptor has a specific molecular tuning curve, which determines the odours it responds to, but these curves overlap to some extent. This is the secret behind the enormous coding capacity found in the olfactory systems of all animals. By combining these spectra many, many different odours can be distinguished by much fewer receptor types. A receptor can be extremely specific and respond only to one or a few odours, while others respond to almost anything.

The neurons carrying the receptors are all embedded in the main olfactory epithelium, the olfactory mucosa. There they swim in a layer of mucus, what we informally call snot. The mucosa is to some extent divided into zones, where a certain type of neuron is more probable to occur in a certain zone. From the mucosa all smell neurons send their axons to the main olfactory bulb. By using neurogenetics, my colleague Peter Mombaerts, Director at the Max Planck Research Unit for Neurogenetics, was first to show that neurons expressing the same receptor usually target two specific sites in the bulb, two glomeruli. Incoming information in this way gets parsed into a spatial code.[3 4] Basically, the mouse nose, its mucosa and its main olfactory bulb thus mirror our olfactory system very closely.

What does the mouse smell with its normal nose? Well, more or less everything of significance.

The vomeronasal organ

The vomeronasal organ (VNO) or Jacobson's organ is a second nose of the mouse and many other animals.[5] It was originally discovered in snakes back in the 1700s by Frederick Ruysch, a Dutch anatomist, but was later rediscovered and described in 1803 by the Danish surgeon Ludwig Jacobson, to whom it owes its name. It's located just above the palate and consists of two cylindrical structures lined with a type of mucosa carrying about 300,000 receptor neurons equipped with three main types of VNO-specific receptors.[6]

The VNO receptors are thought to be mainly involved in detection of pheromones (see below) and other mouse-emitted smells, including those signifying illness. Two of the receptor types bind small, volatile molecules, while the third seems to bind very heavy, waterborne polypeptides like the mouse urinary proteins, or major urinary proteins (MUPs). Despite being involved in pheromone detection, the different types of VNO receptors seem to be equally expressed in males and females, so no sexual dimorphism exists. From the VNO, smell neuron axons target a separate part of the primary olfactory brain, the accessory olfactory bulb.

The cylinders of the VNO are filled with liquid and connected to the nasal cavity via a tube filled with water. VNO-smelling is therefore liquid-based, which means that another spectrum of odours is detected or other mechanisms of delivery to the neurons have to exist. Some odours can reach the organ and its smell neurons on their own, while some need to be bound by MUPs to be able to stimulate and be smelled. Also, some MUPs seem to work as "smells" themselves.

The VNO is activated by a specific behaviour called the flehmen response, which takes place during social interactions or when socially relevant cues are around. Maybe you have witnessed it in a horse, where it's very obvious. The flehmen behaviour opens up the VNO duct for passage and the stimuli can reach neurons inside the organ. See Chapter 3 to learn how dogs exploit it.

The Grueneberg ganglion

At the very tip of the mouse snout, just above the opening of the nostrils, a structure containing some 300–500 neurons was described by Hans Grüneberg in 1973. The architecture of this organ, known as the Grueneberg (or also Grüneberg) ganglion, is quite different to that of a nose. The cilia of the smell neurons are embedded in the skin, but water-soluble stimuli can still reach them. The neurons express specific receptors, where some are similar to what was found in the VNO. From the neurons the axons gather and run together in a special tract to a specific part of the olfactory bulb called the necklace glomeruli.[7]

The function of this organ was long debated. Many scientists thought that it was involved in pup feeding as it developed very early and was located at a very convenient place, close to the mother's mammaries. When Marie-Christine Broillet and her colleagues at the University of Lausanne in Switzerland investigated the function by checking which odours activated the neurons in the organ, they discovered a totally different, somewhat frightening function.[8] As all scientists they started by testing many different relevant odours. There was no response to any known mouse pheromones, or to milk odour or to urine smell.

Another situation where it's good to have quick access and fast response to a stimulus is of course danger. The scientists started investigating different odours connected to danger for a mouse. What they found was that the key stimulus for many of the neurons in the Grueneberg ganglion is a specific molecule emitted by dying mice.[9] They let mice suffocate in CO_2, a method that is used to put pigs to sleep in abattoirs, but that is also known to provide a not so pleasant death. (I really have no idea why it's used commercially beyond the fact that humans don't have to observe the agony of the pigs…)

The smell emitted from the dying mice contained a number of specific molecules that were not present in live mice, nor in mice that were quickly killed. It's a kind of *Schreckstoff*, or alarm pheromone (see also Chapter 5). In a second study, the scientist nailed down the identity of the odour and, interestingly, discovered that it's very closely related to some compounds used by mice to detect predators like cats or foxes. Maybe the organ has a general function in detecting stuff that should make you really scared.

In addition to sensing scary smells, the Grueneberg ganglion also seems to be involved in sensing cold.[10] If this information is somehow integrated with the alarm smell is not really known. It might be an independent sense of its own.

The septal organ

The organ of Masera, or the septal organ, was discovered in 1921 but first described by Rodolfo Masera in 1943. It is a specific small patch of olfactory epithelium that is separated from the main nose by non-smelling epithelium.[11] It's located right in the middle of the nose, at the septum and deep down, where the nose meets the pharynx. The organ contains about 20,000 olfactory smell neurons that express only about ten olfactory receptors.[12 13] Half of the neurons express one specific receptor, MOR256-3, which is the most widely responding receptor so far characterized. This means that it detects many, many very different odours, but it seems like the sensitivity of the neurons in the organ is much higher than in the normal nose. The neurons of the septal organ send their axons to a few glomeruli of the olfactory bulb.

So, what's the septal organ good for? Well, so far no one really knows. One popular suggestion is that it just tells the brain that now some odour is around and prepares it for more detailed information coming in from the main system. An alternative suggestion has been that the septal organ just functions as a mini nose, complementing the main system. An interesting feature of the septal neurons is that they seem to be bi-modal, as they also respond to mechanical stimuli. The neurons could thus monitor both smell and the air speed at the same time and possibly adjust sensitivity accordingly.

In a nutshell, the mouse has four different organs to pick up smells. One is located in the main nose, just like ours, while the others are located in specific areas of their own.

Interestingly, the size of these organs is very different. Ten million neurons in the main nose, 300,000 in the VNO, about 500 in the Grueneberg ganglion and 20,000 in the septal organ. As a result, it seems that some important tasks have been "outsourced" from the main nose to other, smaller organs that serve specific functions: pheromone detection, alarm and, for the septal organ, maybe a general olfactory awareness. In any event, this multitude and diversity of organs just to detect smells testifies to the immense importance of smelling to the mouse and its life.

A life decided by smell

A mouse has four noses and there is a good reason for this. More or less every aspect of the mouse's life is decided by smell in different forms, be it pheromones, predator smells or food cues. The multitude of pheromonal interactions that have been revealed is really staggering, but I will try to unwrap some of the more astounding ones.

A little mouse is more or less a walking factory for different types of pheromones.[14] These can come from urine, but also from the genital glands, tears and saliva. There are primarily two types of pheromones: releaser pheromones and primer pheromones. The releaser ones elicit an immediate response like attraction or aggression, while the primer pheromones change processes in the body over time, often mediated via hormones. Let's look at the releaser pheromones first.

A clear response in male mice is aggression towards an intruder male. When a newcomer enters the territory of a resident male you get a real fight. A female or a castrated male will not get attacked, while a castrate painted with urine from an intact male will get the same aggressive welcome as the donor of the smell. It is clear that it's the smelly compounds present in the male urine that tells the resident male that an intruder has entered the premises. Also, lactating mothers will show aggression towards intruding males, but virgin females will not. Probably a way to defend the pups against the sometimes-murderous intruders. A non-father male will often kill the pups of another male just to get the female ready for his own pups faster. Cruel but efficient…

The aggression from both male and female depends on both volatile and non-volatile compounds in male urine.[15 16] These are detected both by the main nose and by the vomeronasal organ. A complete blend of these different compounds is necessary to get the full aggressive behaviour, which may be triggered to protect your turf, your partner or your offspring from male intruders. This can be seen in many other species, but the mouse has provided unique insight.

Another vital function is to find the mother's nipples at birth. Mouse pups are born totally blind and are absolutely dependent on smelling their way to the nipple. The behaviour they exhibit is very stereotyped and quite similar to an insect zigzagging towards a goal. If the nipple is washed or the main nose disrupted, the pups do not find the nipples of the mother and die from starvation. The whole process seems to involve learning all the way from the mother's womb. The pups learn the smell of the amniotic fluid but also learn smells present in the mother's milk and saliva very quickly.[17] In another small mammal, the rabbit, a specific pheromone has been identified, but in the mouse the system seems to be more flexible and the preference can be changed by the mother's diet. This goes back to what we already wrote about humans in Chapter 2.

The most commonly known releaser effect of pheromones usually involves sex. The same is true in mice. A very intricate odour-driven interaction takes place between male and female. The urine of both male and female are highly attractive to the opposite sex. Both highly volatile compounds and non-volatile major urinary proteins (MUPs) are involved (see above VNO). One MUP in particular was identified by Jane Hurst and a team of researchers from the University of Liverpool in the urine of male mice. Appropriately called darcin after the charming Mr Darcy in Jane Austen's *Pride and Prejudice*, it plays an important role in triggering sexual desire in females.[18]

The opposite sexes are clearly attracted by urinary smells but also by smells emanating from specific glands around the genitals, from saliva and from tears. The smell of male tears really puts the female in a mood for love. All of these smells interact in an intricate way via both the main nose and the vomeronasal organ.

All these types of pheromones seem to function irrespective of the individual. They are general messages regarding sex, aggression and so on.

Who's who?

Another important aspect in the life of a mouse, just like in our own, is to keep track of who is a friend, who is a foe, who is family, who is not, and so on. Most of these tasks are handled by smell recognition in the mouse. Here it's really worth taking a look back at Chapter 2 on humans, where very similar processes are described from our own interactions.

One important aspect in individual recognition is to be able to judge relatedness to some extent. Here, different proteins involved in the immune system play an important role. We already touched on this when we discussed human behaviour earlier. The proteins are again present in the urine of mice and the proportions are individually unique. Neurons both in the main nose and in the VNO can detect these proteins and the smell can guide mice in their mate choice. Not good to choose someone too similar, but maybe not good with someone too different either. Again, the smells decide important choices in the mouse's life.

Another of the more well-known effects of mouse individual recognition is one that we should hope does not apply in humans. The pregnancy block, or so-called Bruce Effect (as you may have already started to suspect, several mouse organs and behaviours are named after the person who discovered them, in this case, Hilda Margaret Bruce, a British zoologist), happens when a pregnant female smells the urine of a male she does not know.[19]

The smell of the stranger's urine causes the female to spontaneously abort her pregnancy, freeing her up to mate with the new, obviously strong male. Why is he perceived as stronger? Quite simply because he had been able to scent mark in the previous owner's territory. The fact that the female can recognize a newcomer means that she carries a clear memory of the smell of her previous mate. The Bruce Effect seems to be dependent on specific MUPs and requires a functional VNO to take place. The main nose doesn't seem to play a role in the interaction.

Priming the body

Primer pheromones change processes in the body in a way that increases the chances for genes to be passed on to the next generation. Several primer pheromones change the female reproductive cycle. Already in 1956, Wesley K. Whitten described the Whitten effect, where the smell of male mice causes group-housed females to synchronize their estrous cycles.[20] This means that they will all be ready to mate when the male is there (also here check out Chapter 2 and the controversial experiments by McClintock). Another smell-driven modulation is that male odour makes young female go into puberty earlier (the Vandenbergh effect – guess who discovered it).[21] Also females can affect each other's cycles. The smell of female odour seems to have the opposite effect as the male Whitten effect. They delay the estrous cycle.

All of these effects seem to be heavily dependent on smells given off from the urine. The types of smells include both volatile compounds but also MUPs and other heavy compounds. Several of the behaviours are totally dependent on a functional VNO, so here the main pathway to the brain an onwards to the hormonal systems seem to pass VNO receptors.

Know thine enemy

There are two important aspects in a mouse's life: reproduction and survival, or maybe the other way around. Survival is paramount, as a dead mouse does little reproduction. Also, here, the sense of smell plays the most important role. Highly specific smell detectors are just aimed at detecting the smell of different predators and both the main nose and the VNO seem to play important roles.

Here it might also be a good place to repeat some more smell terminology. We have discussed pheromones as being specific chemical messengers between individuals of the same species. As mentioned above, there are of course ample cues passing between species as well. The name for these interspecific cues depends on whom they benefit.

If the receiver of the cue benefits, the smell is called a **kairomone**. A typical example is when a prey smells its predator and escapes. If the sender benefits, the smell is called an **allomone**. Examples here would include typical lures to fool a prey or a stink to fend off your enemy. The third category would then be when both parties take advantage of the smell interaction. This is called a **synomone**, and the classical example is a flower odour, where the visiting insect gains nectar, while the flower gets pollinated.

In our story, we see the interaction from the perspective of the mouse in relation to its predators. As the mouse benefits from smelling them, these scents belong to the kairomones. Specific smells from the fox, cat, rat (indeed a mouse predator) and several other species have been identified. Some seem to be detected in a more general way, while some are detected by very specific subsystems. One odour, 2-phenylethylamine, is a true, but general indicator of predators. The smell of this substance truly makes mice run scared.[22]

An example of the more specific predator smelling is a study in the lab of neurologist Thomas Bozza at Northwestern University. His team targeted a small family of receptors, where the function was not really known. In very elegant experiments, Bozza and his collaborators, led by Adam Dewan, could show that a genetic knockout of just one of these receptors stopped the mouse from avoiding certain predator smells. Their study showed that a very specific pathway through the mouse olfactory system functions as an alarm system for the presence of predators. The single gene necessary for encoding the receptor that triggers predator-avoidance behaviour is called TAAR4 (trace amine-associated receptor), which responds to the phenylethylamine (PEA) compound mentioned above, which is, for example, found in cat urine.[23]

So, science shows us that mice are really wary of the smell of cat urine. In practice, this has been put to use for quite some time as cat owners use the soiled kitty litter to deter mice from entering their home by spreading is at strategic places. Sometimes, this practice is not exactly appreciated by the neighbours.

Smelling to survive

Given all the evidence so far, I'm pretty certain that neither Hunca Munca nor Tom Thumb were fooled by the painted plaster. They knew what the material was long before they caught sight of it. A mouse lives in an extreme odour world where all behaviour and many bodily processes are governed by input from the different organs that detect smell.

Food, mother, partners, friends, enemies – everything smells in specific ways to the mouse and allows it to make the right behavioural decisions to maximize its survival and its reproductive success. Just like with the dog, it is more or less impossible for us to imagine all the impressions that can be gained through the different olfactory organs of these tiny creatures that are indeed our house companions all over the world.

Chapter 7

The Best Smeller of Them All: the Moth

Take a kilo of sugar, pour it into the Baltic Sea and stir it all up to create an even concentration. Now take a sip. Do you think you could ever sense such a minimal change in concentration? A moth could. This is exactly what a moth could sense when he's smelling his way towards a female, using her trail of perfume to guide him into her open wings.

In absolute terms, it all comes down to smelling a few molecules in a cubic centimetre of air.[12] To put this into context, our own human threshold stands at around 200 MILLION molecules. And as much as we humans have tried, we still haven't managed to design a detector that comes anywhere near the sensitivity of the moth "nose". Despite moths being a popular olfactory research object since the 19th century.

Back in 1880, Jean-Henri Fabre was one of the first scientists to observe how female moths attracted males from afar. His experiments were simple, but his conclusions were astute.

After excluding vision and hearing as plausible senses, he concluded that odour must play a decisive role:

"There remains the sense of smell. In the domain of our senses, scent, better than anything else, would more or less explain the onrush of the Moths, even though they do not find the bait that allures them until after a certain amount of hesitation. Are there, in point of fact, effluvia similar to what we call odour, effluvia of extreme subtlety, absolutely imperceptible to ourselves and yet capable of impressing a sense of smell better-endowed than ours?"

Fabre was indeed on the right track. The female moth produces an alluring odour to attract the male. This odour is a classic example of a pheromone. Pheromones are chemical compounds that trigger behavioural or physiological responses in another individual of the same species.[3] In many ways, they are comparable to hormones. But while hormones work as chemical messengers within a single individual, pheromones transmit signals between individuals within a single species.

Finding the female

The male moth smells the pheromone – and all other odours – almost exclusively with his antennae. Sitting on the top of a moth's head, the antennae hold a vast number of microscopic smell hairs, known as sensilla. Up to 100,000 of them. This is true of both the male and female moth. Each individual hair, or sensillium, contains a few smell neurons, each one able to detect a specific spectrum of odour molecules. In the male, the vast majority of these neurons are specialized in locating the female pheromone.[4]

Imagine each of these hairs as a mini human nose. Confined in its own environment, each sensillum can regulate the surrounding chemistry of the neurons inside. The neurons themselves are swimming in a viscous fluid that promotes the transfer of molecules into the neurons. As a result, the male moth is at least one million times more sensitive to the perfume of his female than we are to the smells that we are most sensitive to.

The desire to hunt down the female moth also explains the brain structure of the male moth. Almost half of the smell centre of his brain is devoted to the sex pheromone. This centre is divided into a number of little chunks, each of which takes care of just one of those molecules that make up the female perfume.[5]

Why has such an amazing ability to detect minute concentrations evolved? To discover the answers, we first need to dive into the topic of sexual selection – and also consider the predators and parasites that are eavesdropping on all kinds of signals, waiting to find their prey or host.

When so little achieves so much

When a female moth emits her sex pheromone to attract a male, she sends out extremely small amounts of odour molecules. Every hour, this would be the equivalent in weight to one dot on this page. By sending out so little, she achieves two distinct and desirable results. As her smell is so faint, the risk of predators or parasites locating her is kept very low. Her enemies would have to develop extreme sensitivity to be able to sniff her out. It can happen, but it's rare.

For one butterfly species, one enemy has honed its sense of smell to a deadly degree. The hitchhiking egg parasitoid has succeeded in developing such olfactory sensitivity for the pheromone of a butterfly species. Once an egg parasitoid locates a female trail, it tails her, jumps her and hitchhikes with her until she lays her eggs. At that precise moment, the parasitoid jumps off and lays its eggs inside those of the butterfly. Its own grub hatches first and feeds off the organic matter inside the eggs – obviously killing the would-be caterpillar in the process.[6]

Signalling down the genes

Probably even more important in pushing the concentration of the female moth pheromone to extremely low levels is the significance of these kinds of signals in reproduction, especially in sexual signalling. As the female emits such low concentrations, only the males with very sensitive "noses" will find her. This means that there is a constant evolutionary pressure on the male moth to become more and more sensitive. Hence the immense antennae of many male moths.

From an aerodynamic perspective, such huge "wind brakes" at the front of a flying insect are obviously not ideal. Think of them as massive parachutes at the front of a jet liner. The payoff in mating success, however, makes them entirely worthwhile. In many ways, the antennae are similar to a peacock's tail. While this ornate tail certainly hampers the male peacock in his movements, its size makes him a lot more attractive to females than the males with smaller tails.[7, 8]

As the female selects only the very best smellers among the males, she obtains the genes of these super-smellers for her offspring. As a result, her sons are very likely to be extremely good smellers as well. This is known as the "sexy son hypothesis": a female that mates with a male of superior quality will herself get sons carrying genes for this quality.[9] These sons will be better at reproducing or surviving and will transfer the mother's genes more efficiently to future generations. In this way, evolution favours females that put a lot of pressure on the male to find her and, consequently, to mate with her. And it favours males with an even better keen sense of smell.

Weighing up the risks

What about the male? He takes a major risk flying all the way to find the alluring female. During the flight, he is exposed to both birds and bats that specialize in catching flying insects. To counteract this threat, the moth has developed a couple of ears that can detect the bat sound. If he hears the bat approaching, he takes evasive action.[10] However, he will also decide to make a trade-off: if he senses that he's getting close to the female – or close to getting sex – he's much less likely to abort his flight just because a bat is in the vicinity than if he were further away. He's making a trade-off between the chance of being eaten and getting sex.[11]

His enemies try to exploit the moth's obsession with this female perfume. The bola spider, for instance, has developed a particularly cunning hunting method based on exactly this fixation. This spider produces a sticky little ball, the bola, and smears it with the same compounds that the female moth uses to attract a male. The spider then sits on a twig, and lets its bola hang in a thread of silk. When the male moth approaches, fully convinced that a wonderful female is sitting upwind, the spider swings its bola at the moth, who gets stuck, hauled in and quickly devoured.[12] According to the terminology discussed in earlier chapters, the spider's lure is an allomone, as it favours the sender.

The male quest for a female moth is evidently dangerous, time-consuming and energy intensive. But, it's all worth it in the end. The female is a rare and unique resource. When successful, his reward is considerable: his genes are transferred into all the female eggs.

Each to his own

On one level, as a mean of communication, this signalling seems highly efficient. Billions of moth pairs use it to hook up and mate every year. But how exactly do they tell each other apart? We know that there are thousands of moth species in nature. If they all used the same communication channel, male moths would spend most of their time flying towards females of the wrong species. This is of course not the case. The females of each species produce their own, unique composition of pheromone compounds, while the male carries smell neurons exquisitely tuned to exactly these molecules. You could say that each species has its very own olfactory language.

When the first moth pheromone was identified by Nobel laureate Adolf Butenandt in 1959, it was a sensation.[13] He used 500,000 silk moth pheromone glands to get enough substance to identify bombykol, the silk moth pheromone.[14] At the time, the belief was that there are enough chemical compounds for each species to have its own compound. This was later shown to be wrong. Only a limited spectrum of molecules can really be synthesized and used as volatile messengers. To achieve specificity, the female has instead often evolved to produce a very specific blend of compounds. In parallel, the male has evolved an information-processing system that allows him to precisely identify the female of his own species out of a chemical cacophony of many others.[15]

A typical way to study the behaviour of moths is in wind tunnels. These are usually Perspex channels where a slow wind flows through. Upwind, different synthetic mixtures of compounds mimicking the female odour can be introduced. If a male is subsequently introduced downwind, he will perform a stereotypical flight. This flight has evolved into a perfect search pattern; the male flies upwind and at the same time checks his progress in relation to the ground. If he loses contact with the odour plume flowing from the female, he engages in a casting behaviour that gets wider and wider the longer he stays out of the plume. In this way, the chance for him to re-find the plume is maximized. As soon as he smells the perfume again, he rushes upwind. These behavioural steps are reiterated over and over again until he has zoomed in on the female, lands and mates.[16]

Eye-opening, dangerous science

For me, working in the lab, measuring the responses of single smell neurons on the moth antennae was an eye-opener to science. The fact that you look at something that no one has ever looked at before – ever – is really quite a sensational feeling. To see something for the first time: this is really the essence of science. Researching sex pheromones in moths may sound like a hazard-free zone in science, but it's not always harmless and it definitely comes with certain risks. Especially out in the field.

Ernst Priesner, a true pioneer in this field, mysteriously disappeared in July 1994. He was an avid field researcher, always working, all over Europe. He went missing while out on a field trip, setting and checking the insect traps in the alpine region around Garmisch-Partenkirchen, Germany. An Austrian biologist, Priesner specialized in physiology, biochemistry and biophysics of insect olfaction, along with the biosynthesis of pheromones. He was also working at the Max Planck Institute for Behavioral Physiology in Seewiesen.

Essentially, he helped lay the foundations for many breakthroughs in this field. As one of the disciples of the German biologist Dietrich Schneider, the de facto founding father of insect olfactory science and the first scientist who attempted to decipher the mechanism behind the insects' powerful sense of smell back in the 1950s.

When Priesner hadn't returned as expected, the local mountain rescue teams were dispatched and search parties sent out, all sadly in vain. He was never found.

Asymmetric tracking and female pressure

Over evolutionary time, new species are born. One way this can happen is that the pheromone changes. The question is, how can this happen? Normally, a communication system should stay very stable over time. Any male or female that changes in either bouquet or preference would risk remaining without a mate. And yet, we see changes going on. A theory explaining how such changes can occur is called asymmetric tracking. It builds on the fact that there is always a male out there for every female.[17]

This means that even if a mutation causes a female to smell a bit different from the norm, the male is so eager to find her that there would always be a few males ready to go for the new perfume. In this way, a section of the population could start producing and responding to a new pheromone and, over time, form a new species.

In Europe, we found one example of how such processes of change can produce many different olfactory languages within the same species. When Swedish turnip moths were compared to those of the same species in France or Bulgaria, it was clear that both males and females were different. A French female could very likely not "speak" to a Bulgarian male and vice versa. The same species occurs south of the Sahara. In Zimbabwe, a big part of the pheromone bouquet had been dropped during evolution. It's very likely that such changes have occurred under geographic isolation and new dialects or languages have formed. In some instances, they are so different that we could start talking about new species.[18]

In a few moth species, the roles of sender and receiver have been reversed. The male produces the odour and the female responds to it. Nevertheless, the risk-taking often remains with the male, as observed in some tropical moth species. There, males form groups hanging in trees and pump out huge structures from the back of their bodies. These structures emit the pheromone. Females are then attracted to the groups of males.[19] [20]

This behaviour can be compared to the lekking behaviour in other animals, where males gather and display for females attracted by their joint action. Females then choose to mate with males according to certain signals or possessions. In the tree-hanging moths, it could be the size of the smelly structures.

Moths and our ecosystem

The sex pheromones are just one aspect of these fascinating creatures. Another is concerned with the moth's role as a pollinator. It's no secret that flowers produce scents to attract the plant's pollinators. In the case of tobacco plants, the pollinator is the hawkmoth. How does the hawkmoth detect the floral scent – and how does that scent affect their behaviour and flower-moth interactions? This was part of some fascinating research, with sensational results.

Our research required genetically silencing the scent of some tobacco plants, to create plants that either emitted a scent or didn't.[21] We then allowed hawkmoths to choose between unscented plants and their scented counterparts, and analysed their flight patterns, approaches and flower contacts. The work was carried out in state-of-the art wind tunnels and in a free-flight tent. Although hawkmoths visited the unscented flowers frequently, showing that they are not invisible to the moths, the moths didn't stay long enough to successfully pollinate these flowers. The scentless flowers matured very few seeds, which reflects the inferior pollination services provided by the moths – despite a similar number of visits.

We discovered that the moths spent an increased probing time at the scented flowers, which increased their pollination success rate. They also collected more nectar per flower visit. The experiments showed that scent was an important consideration for hawkmoths once they got up close to the flowers. This is possibly because the floral scent is known to reflect the potential amount of nectar available in a given flower.

During our series of wind-tunnel tests and free-flight tent experiments, we assessed the neurophysiology, anatomy, and genes and were able to show that the hawkmoth's proboscis is capable of perceiving flower scent. In fact, as the length of the proboscis prevents the olfactory receptors on the antennae from their olfactory perception, it appears that the proboscis effectively combines the functions of a nose and a tongue, sniffing for the scent at close range to the plant, and sucking up the sweet nectar as a reward. To really find the right way into the flower the moth smells with its tongue.[22]

Smelling flowers at close range with their proboscis therefore helps hawkmoths forage efficiently, which is a key service to our ecosystem.

CHAPTER 7

Working on breakthroughs

As well as being pollinators, moths are also among the most devastating pests on our crops. A huge outbreak of the crop-devouring fall armyworm has been sweeping across Africa, wreaking havoc on many different crops. The same holds for different types of corn borers, the cotton boll worm, and so on. More or less every agricultural crop has its moth attacker. It's the same with our forests, where processionary moths and gypsy moths defoliate many trees. Many of us also encounter small moths in our homes, tucking in to dry goods and our clothes.

For all of these problems, it was thought that pheromones could offer a solution. This has not really worked out as hoped. In some moth species, there has been success with pheromone-based strategies, in others not. The most popular method focuses on mating disruption. This is because when an environment smells too much of pheromones the males get inactivated.

No matings take place. That means no larvae are produced and no crops are damaged. Today, there are also experiments underway to make crops themselves smell like pheromones and produce their own mating disruption. More about this in Chapter 14.

The negative effects of moths are mostly a result of human monocultures and human accidental introduction of species into new environments. In general, moths are very important pollinators and serve as food sources for many birds and mammals. They are also absolutely fascinating to observe at night. Try it yourself sometime with a blacklight and a white sheet. You will make many new acquaintances.

Chapter 8

Even the Smallest Fly

The morning after the party, the wine glasses are still standing around. Hovering over the top, there's a swarm of those pesky little vinegar flies. Inside, a few of them are already dead, drowned in the burgundy. At that moment, you might find it hard to believe, but these tiny creatures buzzing around your kitchen and floating in your wine are giants in the research world. They are one of the world's most important model organisms.

These vinegar flies are the source of so much insight into the basic principles of the sense of smell, from molecular processes to behaviour in the field. For research purposes, they go by their official scientific name, *Drosophila melanogaster*.

But let's return to the kitchen and the wine glasses. Which processes have drawn the flies to the wine? And why? The name of the species is a handy hint. These flies, and many of their relatives, are not fruit flies, but yeast flies. They are highly attracted to fermentation products of a different kind.[1] Vinegar flies prefer yeast that is growing on fruit, and as wine is a product of fermented grapes, it emits the very molecules that will make the fly attracted to the source.

You can test this out yourself and build your own vinegar fly trap to get rid of those annoying flies around your fruit bowl! Although you can easily find special fly traps in many shops and online, it's easy to create one yourself:

In a small glass, pour 10 ml of balsamic vinegar, 90 ml of water and 1–2 drops of detergent (the less smelly the better).

Place the trap close to where the flies are hanging out.

After a few hours, you will find most of them dead – drowned at the bottom of the glass.

How does it work? The smell of balsamic vinegar is irresistible to these flies. Normally, these creatures are so light that they can land, balance and survive on the surface tension of the water. As soon you add detergent to the mix, however, you remove the surface tension. And the small flies fall into the liquid. Devious perhaps, but extremely effective.

In my own research, I have gone considerably further. Having dissected the very odours that attract these flies, my studies have determined that there are about five different types of odour molecules that form an almost irresistible bouquet for this fly. But first, let's delve into why I chose the fly.

CHAPTER 8

The perfect model

The flies may be tiny, but as an organism they have a kind of simple complexity. What makes these flies so attractive to scientists? In the sense of olfaction, they possess a system that allows us researchers to dissect it more or less neuron by neuron. In addition, their genetic system has been studied and dissected since the early 1900s, and a large number of tools to manipulate the system are available.[2] A last, but very important, aspect is their fast turnover. In just a few months we can have many generations of flies, thereby making them perfect targets for studies of evolution and for genetic manipulation.[3]

As the model species for scientists working on olfaction, the vinegar fly has had its sense of smell dissected more times than any other organism. Every smell neuron that detects odour molecules on the antenna and the palp has been studied extensively in minute detail, including what kind of receptor it expresses, which molecules it detects and which part of the brain it will send its message to.[4]

The fly's antenna holds about 1,200 of these neurons, each expressing one or two of about sixty receptor types. The information they receive is sent to the olfactory part of the brain, the antennal lobe, where the meeting places between input and output form little balls of tissue: glomeruli. Each glomerulus is a functional unit taking part in the processing of a specific spectrum of odour molecules. The spectrum can range from a single type to a much wider group.[5]

When an odour hits the antenna, a functional map is created over the array of glomeruli. It's a map that can then be transferred to higher brain areas where memories are stored, and innate behaviour is coded. Most odours are detected by several receptors, which means the array of glomeruli can be played much like a piano, coding thousands and thousands of odours using only the sixty types of receptors.[6 7]

Just as the mouse relies on more than one nose for olfaction (see Chapter 8), the fly also has a second, distinct organ to smell with: the palps. These surround the mouthparts of the fly and were long thought to have a special function in extremely close-range evaluation of smells.

Recently however, we could show that the palp seems to work more or less as a second antenna and detects odours that attract the fly from afar.[8] Why the sense of smell is located on two different organs still remains an open question.

A memory in the nose

When we look at the tiny proteins responsible for detecting different odour molecules in the nose and antenna, insects stand out. Over evolutionary time they have invented a new component in the system. In our nose the receptor identifies the molecule and sends a signal through the cell that opens ion channels and thereby creates an electrical nerve signal. In insects, the ion channel function has been hooked up directly to the receptor, offering a very fast and safe way to provide the transition from chemical to electrical signals.[9, 10]

Another intriguing fact is that this special architecture seems to provide a short-term memory already out in the antenna. If an extremely weak odour stimulus – too weak to elicit an electrical signal – hits the smell neuron only once, no further reaction takes place. However, if a similar stimulation happens again within a certain timeframe, a reaction is triggered, and a signal goes to the brain. A single hit by a weak stimulus is probably nothing to worry about, but if it's repeated it probably means that there's something to it. One time is no time, but two's a bunch.[11]

More than a lab test

In my research, I have investigated the smell ecology of the fly. In other words, I'm interested in understanding the fly as an animal – and not just as a flying test tube for genetic experiments. Along the way, I've witnessed some interesting new principles emerge. The most important one unlocked a secret to the fly's survival.

We discovered that key odours for survival and reproduction are not coded across that many receptors at all. Expanding a little on the piano analogy, using many receptors would be similar to playing chords. Instead, some odours are detected by a single devoted line – which is like pressing one single piano key. We had known about such principles from sex communication systems, as you saw in Chapter 7 about moths, but not really from anywhere else.

The first striking example that we identified was this fly's geosmin-detecting system. You may not recognize the term geosmin, but humans are also highly sensitive to this odour. At low concentrations, it's the pleasant smell of a freshly ploughed field. At higher levels, it's the smell of corked wine or an old mouldy cellar. In the fly, we found a single individual receptor absolutely specific to this odour – and to nothing else. What's more, this receptor also detects geosmin at extremely low levels.[12] Why?

Triggering survival tactics

In a long series of experiments, we were able to demonstrate that the fly uses this geosmin-detection system to avoid fruit that has gone bad. As mentioned above, these flies feed on yeast on fermenting fruit, so they are constantly searching for fruit at the Goldilocks level of decay – rotting, but not too rotten. If the decay has gone too far, other micro-organisms, such as bacteria and mould fungi, invade the fruit. This makes it deadly to both adult flies and larvae. The fly's reaction to the toxic level of decay is similar to the one you might have on opening your fridge door to find a long-forgotten dinner from weeks before.

We humans also have strong avoidance systems to stop us from eating food that has gone bad (See Chapter 2).

How does the avoidance process play out in the fly? This is where the geosmin comes into play. It's the smell of corked wine, but also the smell of toxic bacteria and mould fungi.

When neurons on the fly's antenna detect this bad odour, a signal is sent upwards in the brain, directly into a channel that is singularly devoted to this situation: an ecologically labelled line. The information is hardly integrated with any other additional input. Instead, it is kept apart as a unique alarm signal that says, "Don't go there! Stay away!"

Performing our experiments using light- or temperature-sensitive ion channels introduced in the geosmin-smelling neurons, we demonstrated that we could stop the flies being attracted to otherwise super-attractive food sources – by activating the neurons artificially by light or heat. In the same way, by using mutant flies lacking the specific geosmin receptor or by using mould lacking the enzyme to produce geosmin, we could abolish the stop signal, which meant that the flies ate the bad stuff and died.

Doomed, not doomed?

Another deadly threat to flies (or rather their larvae) is the parasitic wasp. These tiny insects inject their eggs into the fly larva, which becomes a living food storage for the growing wasp larva. Technically, as the wasp larva kills the fly larva, it is not a true parasite, as they would live off the host but not kill it off. In fact, it is a parasitoid that clearly causes the demise of its host.

In nature, parasitoids have infected up to 80 per cent of fly larvae. They are doomed. As you can imagine, the pressure on the fly to develop countermeasures is therefore huge. We discovered one that is based on smell. The fly has developed another ecologically labelled line through its olfactory system that specifically detects the sex pheromone of the parasitoid.

When the fly smells the parasitoid odour it flies away. Amazingly, the larva has the same reaction. When its tiny smell organ picks up the same odour, it starts wiggling and wriggling and moving away, in the process making it hard for the parasitoid female to inject her egg.

Conventional wisdom has it that olfactory neurons in antennae and noses express one single odour-detecting receptor. In the adult fly, the parasitoid detector has broken this rule. We discovered that two different receptors, each detecting the pheromone odours of two different fierce enemies, are found in the same neuron. This points to an interesting fact. One and the same channel can be used to detect odours that have the same negative significance to the fly. The fly does not care which species, whether it's A or B, that is killing its larvae. It just knows that it had better quickly buzz off somewhere else.[13]

The female side

Have all the labelled lines evolved to detect bad odours? Not by any means. In the female fly, for instance, we discovered a specific line for a very different, but also crucial odour.

Alongside the hugely important ability to sniff out dangers to ensure survival, we have another vital function: reproduction. In the case of the female fly, this means finding a suitable place to lay her eggs. As the larvae cannot move very far, their survival is totally dependent on the mother's choice of egg-laying location.

For this particular situation, we found that a specific line has developed to detect the odour of citrus fruit. Citrus is extremely attractive to vinegar flies overall (as you can easily find out yourself in the kitchen), but especially to gravid females – the ones carrying eggs. In this case, the limonene of the lemon and the valencene of the orange are the active odours.[14]

Some unexpected and serendipitous events in the lab provided the confirmatory evidence we needed. One of our students had been working on a general project to investigate attractiveness, using a device where the flies move in little tubes against the wind. One day she came to one of the group leaders in the laboratory with an unusual request. She asked, or pretty much begged, to be allowed not to test limonene any more. The tube was always filling up with eggs from the females, she complained, and it was a time-consuming pain to clean. The explanation was of course that the limonene had triggered the egg-laying labelled line in the females in the tube – and they naturally, and instantly, dumped all their eggs.

Taking evolution into account

What's important to bear in mind when doing this kind of investigation is the evolutionary aspect. In the case of a citrus-specific olfactory channel, we started wondering if it all made sense from this point of view. The vinegar fly has evolved in the African landscape. Citrus fruits come from Asia. It would have been very hard for the fly to evolve a system to detect something that did not exist in its own surroundings (as it evolved).

This fact triggered us to start a major import venture of strange African fruits, one stranger than the other. We collected odour from them all. In the end, we found an African fruit that looks and smells like an orange but is otherwise totally unrelated: the African squirrel nutmeg. It is entirely possible that the fly evolved its preference for citrus-like odours on this fruit. This meant it was already pre-adapted to like all types of citrus fruit when it invaded the rest of the world as a human commensal.

Several other labelled lines help the fly to navigate around the odour landscape, to find what it needs and to avoid enemies and toxins. As already mentioned, specific smell lines are often involved in sexual communication. As you want to find your mate among your own species and before anyone else snaps him or her up, these lines have to be both sensitive and specific. This is true for many different insect types. In the vinegar fly, these mating lines also exist and are crucial for the interaction between the sexes.

The male has developed an especially effective way to ensure that he will sire the larvae produced after mating. During the act, he transfers a specific smell to the female. This smell is a total turn-off for other males – and so ensures that no male will try to mate with the female again too soon. Too soon would be when another male's sperm could replace the sperm of the first male before it had time to do its job.[15]

These examples show that smells can be detected and coded in different ways by the olfactory system. Most general odours are picked up by multiple different channels and form a combinatorial pattern in the system. A few, often of extremely high significance for survival and/or reproduction, are, however, detected and processed by totally devoted, ecologically labelled lines.

A deeper understanding

If you want to understand evolution, and specifically how an animal has adapted to its special way of life, you should really study many related species that have different lifestyles. The fruit flies (which include the vinegar fly) offer such a possibility, particularly as there are more than a thousand more-or-less related species. Many of them are attracted to rotting fruit, just like the vinegar fly, but there are also others that fill many other niches in nature.

We find them in a very wide spectrum of fruit and vegetables, but also in the gills of land-living crabs and even in the guano of vampire bats. In my career, two interesting systems stand out. One lives in a fruit that kills all other flies, while the other chooses fresh fruit over the rotting one.

On the Seychelles, an island nation in the Indian Ocean around 1,100 miles from the Tanzanian coast, a specific species of fruit fly can be found: the aptly named *Drosophila sechellia*. This species piqued our interest because of its diet. It feeds almost exclusively on one fruit: the morinda, or noni fruit, which grows on bushes and small trees. It has a very interesting bouquet. It smells like a mixture of pineapple and Gorgonzola cheese. This smell is the product of a high production of specific esters and an even higher content of acids.

Strangely enough, the acid content is so high that it kills most other fly species. The *sechellia* fly has, however, become a specialist that depends on this food that is toxic to others. By the way, the fruit and its juice are also considered highly beneficial for many different human ailments.

What was interesting was that when we took a closer look at the antennae of the *sechellia* fly, we noticed that the number of neurons detecting the specific compounds emanating from morinda fruit had not only increased in number but had also changed their specificity somewhat compared to the vinegar fly. Inside the brain, something else was going on as well. The glomeruli taking care of the information on these odours had increased in size. The whole smelling system had been focused to detect the sole food source of the fly. The nose and the brain had become a pineapple-Gorgonzola super detector.[16] [17]

Why would such a specialization occur? First, the morinda fruit is very common and offers a year-round food supply. Second, we discovered that the *sechellia* fly is totally dependent on eating morinda to produce eggs. When we further investigated the system, it turned out that the *sechellia* fly has a Parkinsonian-like mutation, which has resulted in very low production of the neurotransmitter dopamine.[18]

The effect on the egg production is major. Flies grown on synthetic diet hardly produced any eggs at all. The morinda fruit, on the other hand, contains a very high level of L-dopa, a compound that can be used to overcome the bad dopamine-inhibiting mutation. Flies fed on morinda-containing diet produced many eggs. During the course of its evolution, the fly had started to self-medicate against infertility by eating morinda and fought against a lack in a vital system. In parallel, to be able to eat a fruit that should have been toxic to it, the fly had to also develop a very strong tolerance to acids along the way. How these different processes are connected, and which one caused the other is still unclear. It's the dilemma of the chicken and egg all over again.

New developments, new dangers

Another fruit fly species that's been in the news recently, one that has raised concerns all over the world, is the spotted-wing fruit fly, *Drosophila suzukii*. Why is it on our radar right now? It has invaded a new niche territory, one where it directly competes with us humans.

A native to Southeast Asia, the spotted-wing fruit fly has now spread, thanks to the global fruit trade, to North and South America, Africa and Europe. Today, it attacks high-value crops all over the world. It has also entered our vineyards. The problem is, the female of the species is attracted to fresh fruit, not to rotting ones like most of its relatives. And it has a fondness for soft-shelled berries, including strawberries, blueberries and raspberries, as well as cherries.[19 20]

The female uses these berries as a place to lay her eggs. To be able to find the fruit, this fly has again evolved its sense of smell. It is able to detect the specific compounds emitted by non-fermenting fruit, and also by the green leaves surrounding the berries and fruits. We know this as the flies attack them when they are still on the bush or tree. This behaviour makes it so much more damaging to crops. In contrast, other species prefer to insert their tiny eggs into fruit that is already decaying – and therefore no longer of value to the food industry.

How is it that the spotted-wing fruit fly is the only species to sniff out ripening fruit? It's the only one that can pierce the tougher resistance of this type of fruit – the resistance this fly meets in ripening fruit is far stronger than any decaying fruit could give. Its secret weapon lies in an organ specifically for this operation. The female of this fly has evolved an egg-laying apparatus, an ovipositor, that looks remarkably like a miniature saw. She can use it to cut through the outer shell of the fruit, and subsequently deposit her eggs.[21] For the fruit industry, the economic impact of this invasive saw-wielding pest has been devastating. It has caused billions of dollars in damage in berry and fruit crops.

These are just a few examples of the many, many lifestyles that we encounter when studying different species of fruit flies. All the specific adaptations they exhibit offer us unique possibilities to understand how specific environments and food choices create specific demands on the sense of smell, which consequently evolves in different directions. If you live on yeast, your nose has to pick up indicative odours of fermentation. If you live on morinda, you have to smell the specific odours of that particular fruit. If you live in the guano of vampire bats, we can only imagine what your nose should be tuned to.

Chapter 9

Mosquitoes: Smelling Blood

There are many dangerous animals on our globe. Usually we think of tigers, sharks, crocodiles and other big predators. The fact is that the biggest killer of humans is a tiny insect, the malaria mosquito, or rather something even smaller, the unicellular *Plasmodium* parasite transferred by the mosquito to humans, causing malaria.

According to the World Health Organization, in 2018 an estimated 230 million people suffered from the disease. Out of these, over 400,000 died. Sadly, about two thirds of the deaths occurred among children under the age of five.[1] Other mosquito-borne diseases like yellow fever, dengue, chikungunya and the Zika virus take another 300,000 lives. To put this into some perspective, the deadliest land-living, large wild animal is the hippopotamus, which takes about 500 human lives in Africa each year. In the longer perspective, since the dawn of mankind, mosquito-transmitted diseases are thought to have killed around 54 billion people, that's around half of the men and women who have ever walked on Earth!

The malaria mosquito – or mosquitoes, as different species transfer the disease in different areas – all belong to the genus *Anopheles*. I will restrict my story about smelling in mosquitoes to these species and to their deadly and ingenious part in the disease chain.

A chain of events

To understand how malaria works, a little background information is called for.[2] Malaria has two hosts and goes through cycles of development in both the *Anopheles* mosquito and in humans. Once in the human bloodstream, a specific stage of the *Plasmodium* parasite – a sporozoite – travels to the liver. These parasites first enter the liver cells, grow and multiply, and start to move out into red blood cells, where they go through different stages, releasing offspring parasites in the process, which in turn invade new red blood cells and start new cycles. It's these blood-based parasites that make people feel the malaria symptoms.

During these cycles in the red blood cells, specific forms – gametocytes – start appearing. These occur in male and female forms. When they enter a female mosquito after she has taken a human blood meal, they start to mate in her stomach. There, again, they start a cycle, and grow and multiply. After a period of 10–18 days, the *Plasmodium* parasite migrates to the salivary glands of the mosquito in the form of sporozoites. When the female mosquito bites the next human to get a new blood meal, she injects saliva to prevent the blood from coagulating, which would end up clogging her mouthparts. With the saliva, the sporozoites enter the human blood stream, travel to the liver and start all over again.

In fact, the mosquito also suffers from the load of parasites and has evolved different counter measures to fight them off, but this is another story.

Optimizing opportunities

From the mosquito's perspective, life is short and has to be optimized in different ways. For the male, the objectives are to survive, eat and mate. For the female, the same things hold true, but she also has to find a suitable place to lay her eggs. It turns out that many of these tasks are dependent on odour information and therefore inherently on the mosquito's sense of smell. In this chapter we will take a look at the life of the malaria mosquito and all the different aspects involving smell and smelling.

Like more or less all insects, the mosquito uses its antennae to smell. In addition, it smells CO_2, and, just like the moth in Chapter 7, also has some olfactory neurons present on its very mouthparts, the palps. The mosquitoes, and especially the male, also uses their antenna for hearing, which we will return to later on.

Smelling the flowers

When we think of mosquitoes, we associate blood-sucking, annoying little insects. The fact is, most mosquito feeding happens elsewhere. Male mosquitoes only feed on nectar from flowers, and females do so as well to gain energy both before and after a blood meal. This flower-feeding practice turns out to be quite specific and depends on the aroma of the flowers. Until recently, the flower species visited by the mosquitoes was not really well known.[3]

By using a very smart strategy, my colleagues at the International Centre for Insect Physiology and Ecology, (ICIPE), in Nairobi, Kenya, could for the first time really show which flowers the mosquitoes actually use for feeding. They collected mosquitoes in the field. In a simple first test they established that the mosquito under investigation contained fructose, which showed it had eaten nectar. Then they proceeded and did DNA barcoding of the stomach content of the mosquito and could establish exactly which plants it had taken nectar meals from. They didn't stop there.[4]

By collecting the flower scents from all the flowers identified in the DNA barcoding and testing these in electrophysiological experiments on the mosquito antennae the researchers could establish exactly which molecules – in other words which smells – the mosquitoes used to find the food. It turned out that two smells were present in all the samples and seemed to be crucial for mosquito attraction overall. Additionally, there were a few other smells that could indicate exactly which species the nectar was taken from. It therefore seems like the mosquito has developed a general sense for the smell of flowers that provide a good meal.

We will also see the same compounds popping up in other contexts later on in the chapter.

As we saw above, mosquitoes also suffer from being infected by the malaria parasite. Can they do something about it? Well, it seems that some species self-medicate by changing their source of nectar meals to flowers that provide some compounds with anti-malaria effects. By eating the right stuff, they decrease the parasite load and can produce more offspring.[5]

In general, the malaria mosquito has developed a very keen sense of smell to determine the identity of flowers that will provide ample nectar rewards or that will supply nectar with medicinal properties. The flowers seem to share some common smell characteristics that generally spells "nectar" to the mosquito.

Smelling the blood

Many mosquito species have evolved to take blood meals from different animals. This is likely an adaptation to get access to proteins and especially to nitrogen in an environment that is very poor in these nutrients. Blood is also quite easy to digest and provides an instant kick to the system. The malaria mosquito is more or less a specialist on humans. Other species go for birds, cows, reptiles and so on, and the choice clearly depends on the smell.

It is only the female mosquito that takes blood meals, and she does so as a preparation for egg-laying. The nutrients in the blood are necessary to produce the full egg load, but also to extend her life expectancy and to generally increase the amount of energy available for flight.

How does she find you for that crucial bite? Well, as discussed in Chapter 2, we humans really do send out a lot of molecules from our skin, both on our own account and thanks to all the microbes inhabiting our surface. We also let out telltale molecules every time we breathe.[6]

Let's take a look at all these different smells. Some years ago, Dutch researchers realized that malaria mosquitoes were attracted to the smell of sweaty feet. Yummy, carboxylic acids!

They performed wind-tunnel experiments and could show that the mosquitoes really liked this human odour. They wanted to compare it to something else of similar calibre and remembered the smell of Limburger cheese, which indeed reminds you directly of old socks. And lo and behold, the mosquitoes were just as attracted to the cheese as to the feet of the experimenters.[7]

Now, the sweaty feet turned out to be only a small part of the story and reveal how important it is to take natural concentrations and proportions into account when working on smell-driven behaviour.

Rickard Ignell and his co-workers in Sweden really wanted to find out what the natural smell that attracts malaria mosquitoes to humans looks like. By combining chemistry, behaviour, physiology and fieldwork they managed to nail down a complex blend that mimics a human at natural concentrations. The blend contains old suspects like 1-octen-3-ol (the smell of mushrooms) and several aldehydes and monoterpenes. Interestingly, some of the odours identified as attractive in the flowers mentioned above, reoccurred in the human smell.

In addition to all these skin-produced smells, the breath contains molecules that seem to increase the attractiveness further. Each time we exhale we emit a cloud of CO_2 and also of acetone. Both these compounds have a boosting effect to bring in the mozzies. The fact that acetone is an attractant might also explain the fact that diabetics seem to be highly attractive to mosquitoes.[8] Acetone levels are often higher in the breath of people with diabetes.

Bringing all these different studies together gives a pretty complicated picture of the human odour cues used by a female malaria mosquito to find us. Carboxylic acids (the feet...), ammonia, aldehydes, monoterpenes and mushroom odour (the skin) and CO_2 and acetone from the breath. And all in the right concentrations and proportions.

CHAPTER 9

Are we different?

As we saw in Chapter 2, how humans differ in the way they smell both depends on genetics but also on skin microbiome and diet. Probably these factors are also interconnected. In parallel, we all know of stories of someone who is extremely attractive to mosquitoes, while the husband or the friend doesn't get bitten at all. Is there any scientific background to these claims?

Willem Takken and his group in the Netherlands tested a group of people and compared the attractiveness of their emitted odour to malaria mosquitoes. There were clear differences between the twenty-seven people who took part in the study. Some were significantly more attractive than others. The smells that are actually responsible for the differences still remain to be identified.

As there are differences between people, you could also expect that these differences would be inherited – that they would have a genetic component. To figure out these relationships James Logan and John Pickett performed a study where they compared identical to non-identical twins. Their results showed that there is indeed an inherited, odour-based propensity to be bitten by mosquitoes.[9] Their pool of subjects wasn't very big – eighteen identical twin pairs and nineteen non-identical – but the results are still quite clear. There seem to be a clear genetic component to the human odour profile that seems to have a heritability that can be compared to the ones found for tallness and IQ.

There are thus clear individual differences in how attractive we humans are to mosquitoes. Are there also other factors that might influence how prone we are to be bitten? Above, I mentioned that the odour of diabetics tends to be more attractive to mosquitoes. Another study clearly shows that pregnant women, sleeping under bed nets in the Gambia were about twice as attractive to malaria mosquitoes as non-pregnant women. Again, the exact odours mediating this difference are not known.[10]

Another, really interesting difference in mosquito attractiveness was found when scientists compared people with active malaria to others.[11] In a very well-designed experiment, twelve groups, each consisting of three western Kenyan children, were used. Western Kenya is known as a hotspot for malaria. Out of the three children in each group, one was uninfected by *Plasmodium*, one was naturally infected with the non-infective, asexual stage and one carried the sexual gametocytes that can transmit the disease. When comparing these three conditions, it turned out that the smell of the children carrying the infectious stage was twice as attractive to other malaria mosquitoes in comparison to the smell of the others. After the first experiment the infected children were provided with an anti-malaria drug, which totally abolished the differences in mosquito choice. This result shows that it was indeed the presence of the infectious stage of the parasite that increased the attractiveness of the children.[12]

Several studies have shown that people carrying the gametocyte stage of the malaria parasite increase their odour secretion of certain aldehydes and terpenes both from their body surface and in their breath. My colleague Rickard Ignell and his collaborators have also shown that a specific compound called HMBPP (too long to spell it out here...) is produced by the gametocytes. This metabolite acts as a strong feeding stimulant for the female mosquito but also induces increased production of attractive odours from red blood cells.[13]

This is a very interesting result, as it shows that the parasite somehow manipulates its host to facilitate its spread. By making the humans carrying the "seeds" of new infections more attractive to the mosquitoes, the chance to get transferred increases dramatically.

Another, often discussed background to differences in the chance to get bitten by mosquitoes is the diet you keep. Many suggestions circulate: "eat garlic and you will not be bitten!" or "eat vitamin B". Maybe the well-meaning friends offering this advice have mixed up mosquitoes with vampires, as there have been absolutely no effects shown by consuming neither garlic nor the vitamin.

What has been shown (unfortunately...) is that beer consumption increases the attractiveness of men. In experiments in Burkina Faso, around forty men were offered either one litre of the local beer, *dolo*, or a litre of water. After fifteen minutes, Thierry Lefèvre and his collaborators measured how many mosquitoes were attracted to the subjects.[14] The beer drinkers were clearly more attractive, despite having similar body temperatures and exhaled CO_2 levels.

The difference must reside in the odour emitted. The scientist concluded: "Beyond this coincidental side effect of beer consumption, mosquitoes may have evolved preferences for people who recently consumed beer – possibly due to reduced host defensive behaviours or highly nutritious blood meals. This hypothesis is appealing but requires further investigations."

Smelling with the tongue

The task of a little mosquito female is quite daunting. First, she has to travel through the air towards the attractive victim, avoiding birds, bats and human hands on the way. But even if she reaches the skin and manages to land without raising suspicion, she still needs to target a suitable blood vessel.

In collaboration with the lab of Hyung Wook Kwon in Seoul, Korea we discovered a new olfactory pathway that seems to be involved in the last steps of finding blood. When the female mosquito has landed, she penetrates our skin with a structure called the stylet, which looks more or less like a tiny injection needle. What we found was that on the tip of this little needle there were a few tiny structures reminiscent of the smell hairs – the sensilla

– that we find on the antenna. We then looked at the genes expressed in the structure and found two olfactory receptors there. When we tested which odours these receptors recognized we found several given off by blood, among them the mushroomy odour discussed above.[15]

By interfering with the RNA transcription, we could make the little noses on the stylet non-functional. This made the female much slower in finding her way to a suitable blood vessel under the skin of a mouse. It seems that the tiny nose present on the tip of the stylet helps the female in the last stages in her search for blood. This very nicely parallels what we saw in Chapter 7, where the moth's long tongue carries some small, smelling hairs on its tip, helping it to find the way to hidden flower nectaries.

Where to lay your eggs?

Much of what we've seen so far has focused on finding optimal places to feed. After feeding and mating, one important task remains for the female: to find a suitable place to lay her eggs. As the malaria mosquito larvae typically live in small puddles of water, the female's choice is extremely important. It will completely decide the fate of her offspring.

The larvae feed on different kinds of organic matter originating from different parts of the environment. These include plant, insect and crustacean breakdown products as well as micro-organisms such as algae, protozoa and bacteria. The female also seems to especially search out pollen from maize and sugarcane.[16] In the food items mentioned before, the nitrogen content is quite low. In the pollen, on the contrary, the larvae find an excellent source of this limited nutrient. Scientists have even established a connection between the establishment of maize fields in Ethiopia and an increase in the number of malaria infections.

All of the different food sources for larvae smell. The smell of decomposing grass and its pollen has been shown to attract gravid females. Also, the specific odours given off by maize and sugarcane pollen smell very good. Recently, the first microbial odours attractive to gravid females were identified. All of these odours are of course of great applied interest as they could form the basis of traps aimed at capturing the egg-laden females.

Beyond the positive odours, the female mosquito is very likely able to detect the smell of already present larvae and of predators. The cues preventing overcrowding seem to be sulphuric compounds similar to what we saw ocean birds using to find food in Chapter 4 and some flowers use to achieve unrewarded pollinations in Chapter 13. In some other mosquito species, the female deposits a pheromone that attracts other females as well, but this has never been shown in the malaria mosquito.

A smell-driven blood sucker

From all of these examples it is clear that smell for a mosquito is absolutely decisive in driving its behaviour and in deciding its choices. Once the antenna is hit by some telltale molecules of a human blood source, the female will take flight and try to reach the skin of the victim. If the smell contains different proportions between molecules or has some other types of molecules added to it, it might signify a non-suitable host and the female will abort her approach. We therefore know quite a lot about how mosquitoes are attracted to us humans, but still we don't know enough to really stop them. In most cases we resort to an artificial chemical discovered in the 50s. Science takes time to find its way to application. More about this in Chapter 14.

Chapter 10

Bark Beetles: Killers of Dinosaurs

Is it possible for an animal just a few millimetres long and weighing a few milligrams to kill another organism the size of a large dinosaur? It sounds implausible, but this is actually what happens when bark beetles take down the trees in our forests. Well-coordinated attacks by tens of thousands of beetles are killing 100-year-old pines, spruces, elms and other tree species within a matter of weeks, leaving ghost forests in their wake.

These aren't isolated attacks. We see them spreading over the whole northern hemisphere, and their scale is unprecedented. In Canada, the devastated areas of dead pines are visible from the moon – as a large brown belt spreading from the west coast inwards. Millions of cubic meters of wood have already been destroyed and long-standing ecosystems have been altered forever.[1] In large parts of the world, the elm has more or less been driven to extinction by the Dutch elm disease, a fungal attack spread by elm bark beetles.[2] In present-day Europe, the spruce is facing a similar fate as millions of trees succumb to increasing populations and infestations of spruce bark beetles,[3] even if the spruce shows considerably stronger resilience than the elm.

So many, but so unique

Bark beetles of every kind – and there are thousands of species, with at least one for every species of tree – are very much driven by smell.[4] They find each other and their host trees by detecting different types of molecules with their tiny antennae. These miniscule structures make clear just how much the architecture of an insect's "nose" is influenced by its lifestyle. It's the reason why moths have giant, feather-like antennae, as you read in Chapter 7. This is only possible because, unlike bark beetles, moths are faced with few constraints on the size of their smell organs. Since they spend their lives flying around in open space, they just need to be able to fly with them.

In contrast, bark beetles, as their name suggests, have to bore in under the bark of trees and burrow their way through narrow tunnels inside. This makes moth-like antennae impossible. Evolution has therefore provided the bark beetles with small, club-like antennae that can be folded into grooves beside the head. This ingenious solution ensures that the antennae are protected while the beetle is crawling through tight spaces, but also easily extended to become efficient detectors when the beetles move out in the open. This expandable type is just one of many forms of antennae that have evolved across the insect world. It's striking just how many there actually are, as any astute observant of the external morphology of insects will confirm.

If a single spruce bark beetle were to try to attack a standing tree it would indeed be like an ant trying to kill an elephant. Coniferous trees have extremely efficient defence mechanisms. As soon as an insect tries to penetrate their bark, they send out a stream of deadly resin. This amber, both sticky and full of poisonous substances, traps the attacker in a lethal syrup.[5] This line of defence is the reason why we find so many ancient insect fossils. These creatures were caught in the defensive resin flow millions of years ago, to be perfectly preserved as a snapshot of life at the time. Whether we'll ever be able to extract genetic material to clone them is another question entirely – despite what *Jurassic Park* would have you believe.

How, then, can this resin defence in trees be overcome? Beetles have found three main lines of attack. First, they have established a chemical communication system that allows them to coordinate their attack in time and space. Second, they have developed an ability to select vulnerable trees – the ones with a somewhat weaker ability to produce large amounts of resin. And third, the beetle carries a "secret" biological weapon.[6] Let us look at these three lines one by one.

Communicating in numbers

How do spruce bark beetles coordinate their attack on a big tree and, more importantly, how do they know when the tree is fully occupied? To understand these questions, we first need to look at the progression of an attack and the chemical signals, pheromones, involved. A tree is first attacked by a pioneer beetle – a male. He bores a small hole in the bark and starts calling for a female. His main signal is a combination of two odour components that basically say, "Come, come, mate with me and help me take over this tree."[7, 8]

If the male is lucky, and not flushed out and killed by resin flow, he might attract a female. They mate and, together, start emitting even more aggregation pheromone, the type of pheromone that attracts. More beetles arrive, bore into the bark, where they then mate. And so, the cycle continues. In the end, the beetle population gets so big that the tree is in no position to defend itself against the mass assault any more. It's doomed.

After the beetles have mated the females start digging channels through the energy-rich phloem of the tree. This fibrous layer, located just under the bark, is basically a complex transport system for nutrients and therefore responsible for the tree's growth and survival. Along the sides of the channel, one female lays up to eighty eggs. As she does so, the male changes his pheromone production to two other smelly compounds. In low concentrations these are also attractive, but when they reach a certain threshold, they turn into a stop signal. In the final stages of tree colonization, the beetles also start sending out a direct stop signal: "Don't come here – it's full. Go to the next tree!"[9]

When the eggs hatch, tiny larvae emerge and eat their way perpendicularly from the mother tunnel. Their routes out create the beautifully elegant but deadly spiral patterns that have given the European spruce bark beetle its Latin and German names: *Ips typographus* or the *Buchdrucker*, which means "letterpress printer" in English.

To sum up, the chemical communication has four stages. First, the male sends out an attracting scent, or aggregation pheromone, primarily to bring in females, but also to attract other beetles in general. Second, males and females send out the same signals, but coupled they become even stronger. Third, after the egg-laying, the male beetle starts producing a pheromone with a dual function. It's attracting until it reaches a certain level. Then it becomes a stop signal. Fourth, when the tree is full, the beetles emit the ultimate stop signal to thwart further attack.[10] How the beetles know when a tree is full remains a mystery.

A vulnerable prey

Going for trees with weaker defence mechanisms is a key skill in beetles. In lower densities of beetle population, this behaviour very likely plays an important role, as fewer partners in crime are available. At peak densities, these insects seem to ravage everything around them, irrespective of quality – and sometimes even of species. When a selection does take place, a beetle will use its sense of smell and taste to judge the suitability and vitality of a tree. By comparing the levels of different tree-emitted odours, it's very likely possible for these tiny insects to determine how well the tree will be able to defend itself.

What, then, can determine the vitality of a tree? Each tree has its own genotype. Embedded in this individuality, the ability to defend itself will vary. Some individuals are stronger and can produce a heavier resin flow with stronger defensive activity. The general physiological state of a tree is also important.[11] A lack of water supply and rising temperatures can weaken a tree. If it is suffering from heat stress after a particularly dry patch, its ability to muster a strong defence is reduced. That's why long, hot and dry summers can make them more vulnerable – and a prime target for beetles.[12]

Powerful weapons

As the spruce bark beetle enters under the bark of a tree, it also introduces a special weapon to fight back against any resistance by the tree. Each beetle carries spores or hyphae of specific fungi, which they then introduce into the tree tissue. The most well-known of these are blue-stain fungi. If they are carried by the attacking beetles, they will colour spruce timber a bluish black (automatically lowering its commercial value significantly).

These fungal species have a specific odour signature that the beetles are expert at detecting, helping them to find and carry it along.[13] Such microbial attacks weaken the tree still further, assisting the beetle in ultimately killing the tree by congesting water conduits. The fungus might well be on the beetle menu too, serving perhaps as food for its larvae. Some studies suggest that the fungus is even better food than the tree's phloem.

Another infamous example of a "collaboration" between beetles and a beetle fungus is found in the culprits behind the spread of the Dutch elm disease.[14] Several species of elm bark beetles attack large elms by boring into their bark to lay eggs. But here the adult beetles also feed in twig crotches. During this feeding, it is highly likely that the lethal fungus is introduced into the tree. This means that only a few beetles are necessary to spread the fungus to many trees and, as a result, a large forest can be devastated in no time. This is exactly what happened in the largest elm forest in southern Sweden. In just a few years, the whole forest was reduced to rows of tree skeletons. Today, the elm is all but wiped out as a tree species in large parts of the world. All because of a tiny beetle carrying even tinier spores and hyphae of an extremely aggressive fungus.

Avoidance tactics

Finding the right tree is one thing. Avoiding trees that cannot be eaten is another. As we discussed for both flies and moths, the sense of smell also contains specific information channels for "bad" smelling substances. The spruce bark beetle has also evolved these kinds of detectors – and uses them to avoid one smell in particular, the odour emitted by birch trees.[15]

When the birch anti-attractant is mixed with the smell of spruce or pine, normally a desirable host for the beetles, the attractiveness of the scent mix goes down drastically. This fact is important knowledge for our forestry and silviculture, as it shows that heterogeneous stands of trees are more resistant to attack and less susceptible to bark beetle infestation than monocultures. This strategic mixing of host and non-host odours to ultimately deter the most damaging tree pests and increase forest resilience is known as semiochemical diversity.

Hunting the hunters

The bark beetles themselves are an important resource for many other animals. Several insects have evolved the ability to pick up the bark beetle pheromone to pinpoint their location. The *Thanasimus* beetle, for instance, has evolved receptors that precisely detect the pheromone produced by these bark beetles, so it can hunt them down as they roam the tree surface.[16][17]

Other bark beetle enemies combine chemical senses with detectors for vibration, which they use to locate larvae under the bark so as to inject them with parasitoid eggs. Other efficient bark-beetle predators are woodpeckers, of course. These birds use their narrow beaks to pry out the beetles and their larvae from under the bark. Another interesting and smell-driven beetle eater is the wild boar. In my own forest in Sweden, I've often seen these animals seemingly dancing around certain trees.

It turns out that the pigs had been attracted to the trees that were heavily infested with bark beetles. The new generation of beetles had dropped to the ground to overwinter, but the pigs had other plans. By smelling their way with their sensitive snout and walking around the trees, they more or less devour the soil and all the bark beetles in it. In the process, the wild boars do a great service to many forest owners.

The beetles and the ecosystem

While bark beetles have an important role to play in the ecosystem, it's clear that these pest insects have a devastating impact on forests all over the world. In the worst-hit areas, they destroy millions of cubic metres of prime timber every year. Thanks also to highly beneficial conditions for the beetles brought on by the current state of climate change, we're experiencing population explosions of these pests and more frequent bark-beetle outbreaks.[18]

Equally, the cascading effect of climate change is increasingly compromising the forest's natural defences, as you read in Chapter 1.

What's disturbing is that a deadly combination of drought, fire and most likely a certain type of bark beetle (*Phloeosinus punctatus*) could be responsible for the deaths of giant sequoias in California, trees that have survived more than 3,000 years.[19] These sequoias are perhaps the largest living organisms on our planet right now. The beetles really are coming for the dinosaurs of our time.

To meet these challenges, many forest owners are exploring management by smell, as you can read in Chapter 14.

Chapter 11

Christmas Island Crabs

Some creatures on Earth are more fascinating and astonishing than others. The same can be said of certain places. I thought, after experiencing Africa's savannah and the Great Barrier Reef, I had seen the true natural wonders of the world. Christmas Island and its original inhabitants proved me wrong. But, let's trace back in time and find out how smell research and serendipity made this small island a second home to me and my family.

In 2002, my PhD student Marcus Stensmyr approached me with an intriguing popular science article about the world's largest land-living arthropod (an insect-crustacean-spider-animal), the robber crab, so-called for their rumoured tendency to steal bright things. They also go by the official term *Birgus latro*.[1] As we were currently working on the evolution of olfaction among insects, he suggested that we should incorporate some land-living crustaceans for comparison.

Today's largest population of robber crabs is found on Christmas Island, a 135-square-kilometre rock in the Indian Ocean, about 350 kilometres south of Java, Indonesia, and 2,600 kilometres northwest of Perth,

Australia. Despite the distance, the island is Australian territory. Nowadays, it's probably best known for the media coverage on the refugee detention centres, and, sadly, for the capsizing of a refugee ship in 2010.

Throughout most of its history, thanks in part to its remote location, this island has escaped much of the usual devastating impact humans or other mammals bring. It has been inhabited mainly by birds, insects and crustaceans. For the latter, it really became a unique ecological experiment, as crustaceans took over many of the ecological niches generally occupied by mammals elsewhere. A population of over 100 million red crabs decompose leaf matter and keep the forest floor clear in the otherwise dense rainforest. Nipper crabs take on the role of specialized predators, while at the very top of the food chain sits the robber crab.[2]

A subject of evolutionary interest

Interestingly, measured in an evolutionary time perspective, the land-living crustaceans entered their terrestrial habitat relatively recently. Not until five million years ago did they start colonizing the forests and beaches. Compare that to their closest relatives, the insects, who can look back over an impressive 400 million years as terrestrial creatures. Why is this information interesting to us as smell scientists?

Well, insects have had an extremely long time to adapt to smell in air. While land-living crustaceans on the other hand, have had just a few million years to do the same. So, the question on our mind was, do these creatures have an olfactory system. If so, how does it compare to the one we know from the insects. As chemical ecologists, we were of course also interested in observing the behaviour of these big creatures.[3]

As the world's largest land-living arthropod, the robber crab can weigh up to five kilos and measure almost a metre across, between its longest legs. At the front, it has a pair of impressive claws, which it uses to ingeniously insert into the eyes of coconuts to pry them open (which is the reason why this creature also goes by the name coconut crab in some circles). The pressure it can exert is twice as powerful as we can manage with our own jaws.[1]

The robber crab has developed very interesting solutions for a life on land. To breathe, it has evolved a heavily vascularized layer of skin under its shield: a branchiostegal lung. In parallel, the last pair of legs has become tiny little "bottle brushes", which are constantly cleaning the surface of the lungs. The previous gills have gained a kidney-like function. The development of these air-breathing lungs means that the adult robber crab is totally terrestrial and will drown if submerged in water.[4]

A strange life

The life of a robber crab still starts off in the ocean, as egg development has to occur in seawater. On a certain night, often characterized by a full moon, the females move to the rocky beaches and drop their eggs into the ocean. This is a risky business for the females, as they would surely drown if they fell into the water. After about four weeks, the eggs hatch into tiny crab larvae, which depend on currents and tides to be taken to a suitable beach. There they crawl ashore and leave the sea forever.

The robber crab is a type of hermit crab, and as such the tiny new crabs search out a suitable shell to make their home, much like many others of its relatives. They keep changing shells, upgrading to a larger, more roomy accommodation each time, until a certain age, when they, in contrast to all other hermit crabs, develop their own armour and live a life independent of scavenging shells. Still, even if the shell is now made from their own body, it doesn't grow any bigger. This means that every year the giant crab has to dig down into the soil and moult. As it sheds the old shell, it builds a new, even bigger one itself. This procedure goes on and on, year in, year out, for the whole of its life, a life that can stretch up to 100 years![5]

Tracking in the jungle

When we started our research on Christmas Island, precious little was known about the life of a robber crab. We started by following a number of crabs to see what they were doing and how they were moving. This was done by fitting some pretty big individuals with a little backpack containing a satellite-tracking device. When the crab moved through the jungle the device communicated with a satellite providing the GPS coordinates of the crab's location. This data was then stored in the backpack.

To collect the information, we had to get close enough to the crab, about ten metres away, to download the data wirelessly. At that distance, we could also hear the built-in pinger when the download had started. As each tracking device came at a cost of about €1,000 and we had no way to follow the crab, each relocation of one of our crabs was an event to celebrate.

Once we had located a crab it took just a few minutes to download the data. After transferring the data to the computer we could observe where the crab had been walking during its days of absence – and we could also get some indications about what it had been up to. Sometimes, the crab had walked up to two kilometres in a week, at others it had simply been sitting absolutely still for the exact same time span.[6]

CHAPTER 11

Sex on the beach

One topic on which we gained new information was robber crab sex and mating. The reproductive biology of the robber crabs had been a mystery for some time. Females had been observed dropping their eggs into the ocean, but what we wanted to know was what had happened before that. It was a complete unknown. When we started our research, we first had to establish two basic facts: How do the crabs meet? How do they mate?

In our tracking experiments we saw how big males could sit still up in the mountain rainforest for weeks at a time. Then, all of a sudden, they would embark on a kilometre-long walk down to the beach. When we followed the same trail, we found the female crab at the beach, often around freshwater caves close to the seashore. There, we observed the very first robber crab matings ever registered (as far as we know). Interestingly, the crabs do it "missionary style", with the male grabbing the female claws and slowly turning her onto her back so the mating can begin.

Afterwards, the timing for egg-laying was not always perfect. This seems to be where the freshwater caves played a role. In anticipation of the moon phase that indicated the right tidal conditions, the mated females gathered in these caves in their hundreds. I experienced one of those rare David Attenborough-type moments when, after lowering myself down by rope into one of these caves, I found the walls totally speckled with egg-laden, glistening females.

Risky meals

Another behaviour that we observed in our tracking experiments was the gathering of hundreds of big crabs around a specific tree. The tree was always a representative of an endemic palm tree species, the Arenga palm. This palm produces berry-like seeds that seem to exhibit an irresistible attraction to crabs. The crabs were indeed capable of predicting the ripening of the fruits by at least a week, an ability that, as we were able to show later, depended on smell cues emitted by the tree. When the time of ripening got closer, we could even see big male crabs venturing up into the palm tree to feast on the fruits.

Being an empirical scientist I (foolishly) decided to investigate what the crabs found so delicious about these particular fruits. I collected a bunch of ripe berries and popped a couple into my mouth. Bad decision! Immediately after the first bite, my whole mouth went numb. It became difficult to breathe. The fruits were clearly toxic.

After a worrying few minutes the symptoms subsided. I regained the sensation in my mouth and tongue. I could relax. However, the ranger who was accompanying us at the time had just one question for my wife. He wanted to know how I had ever reached adulthood...

There's a lesson for scientists in this story. It shows how careful you need to be in your observations. After experiencing the effects of the Arenga berries, I sat down and observed the crab's eating behaviour up close and in more detail. It turned out that the crabs carefully peel away the fleshy part of the berry with their nimble claws. Then, they proceed to crack the nut and happily devour the fat-rich and non-toxic content inside. Smart crab, stupid human…

The crabs were also after another part of the Arenga palm tree. Whenever a tree was blown down, we noticed that the broken trunk would attract large numbers of robber crabs after only a day or two. We mimicked a storm, cut down a tree and split the trunk, in the process exposing the marrow. Again, after a day or two, large numbers of big crabs gathered on the sections of trunk and happily devoured the marrow. After a while, we observed that the crabs seemed to be acting like they were drunk, wallowing around and sometimes falling over.

When we smelled the marrow, it was clear that alcoholic fermentation was going on. The broken tree had become a robber-crab bar!

Do they smell?

When we first arrived on Christmas Island in 2003, we had a very vague idea about both the organism and the environment we were going to study. The whole family, with two children aged four and six in tow, set out on the long journey to Christmas Island in late autumn.

Everything went well, but landing on Christmas Island was interesting to say the least. The landing strip counts as one of the most dangerous in the world and the pilot has only three attempts to land before he would be forced to return to Jakarta to refuel. Fortunately, we made it on the first attempt.

We had permission to perform a number of experiments and to catch some robber crabs. The very first evening we took our run-down Toyota HiLux and drove along the roads. To our amazement, we could pick up giant crabs along the way. However, we were not aware of how strong the crabs were. We had tried to restrain them in buckets covered with the spare wheel of the car. In the morning, the crabs had simply thrown off the 25-kilo wheel and escaped. Luckily, there were many more of them in the rainforest.

Establishing the facts

We had three main experimental strategies to establish the presence of an olfactory sense: behavioural, physiological and morphological investigations. All the experiments were performed in or around the Pink House, a run-down old ranger station out in the Christmas Island rainforest. To determine if the robber crabs indeed used olfaction to find resources, we first observed what they really liked to eat.

Their diet included coconut pulp, Arenga berries and dead red crabs. In the dead of the night, we put out poles with bags attached containing these different smelly objects. In no time, we could see big robber crabs approaching the bait. It was very clear that in the pitch-black tropical night the crabs could smell their way towards the desired food. Fact one established: robber crabs can smell and use their olfactory sense to locate resources.

Next, my PhD student Marcus and I went into the "laboratory", which was really just an empty room with a desk and two chairs. Fortunately, we had brought portable electrophysiological equipment with us, which allowed us to take one antenna from a robber crab and hook it up for measurements of electrical nerve signals after stimulation with odours. Why the antenna? We were used to the insect way of smelling. But crustaceans have two pairs of antennae, so we had to check them both. It turned out that one pair gave a very strong electrical signal when stimulated with both natural and synthetic odours. Fact two established: the robber crab has an olfactory organ located on its second pair of antennae.[7]

For the morphological part, we collected the brains of a number of crabs. Here, we made use of the sad fact that several crabs got run over every night by the big lorries carrying Christmas Island phosphate to be shipped out (see below). Often, we could extract the brain just after the crab had been run over and in this way, we could minimize our own impact on these magnificent animals. The brain was then chemically preserved and brought home to Germany to be studied in advanced microscopes, where we could start searching for fact three: how the robber crab brain had evolved to enable smelling in the air.

A brain for smelling

After returning to our decidedly more advanced home laboratories, we started dissecting the robber crab olfactory system from the antenna and into the brain. The antenna looked somewhat like an insect antenna but was covered with thick hairs called aesthetascs. These contained a huge number of neurons obviously detecting the different odour molecules. To understand how the molecules were identified we looked for the genes coding for possible smell receptors. It turned out that the land-living crabs have kept their ancient, underwater smelling receptors, but somehow adapted the system to function in air.

What about the brain? Here we brought in Steffen Harzsch, a true specialist on crustacean brains. Together, we followed the olfactory pathway into the brain of the robber crab, where a great surprise awaited us. The crab had more or less invested half of its brain just to process smell information. Evolution had pushed the development so far that parts of the brain had been pushed up into the stalks, on the tip of which their eyes seem to balance precariously, to make up the space. Immense numbers of neurons (in comparison to insects) were processing the olfactory information already at early levels in the brain.[8]

All of the initial experiments pointed us in the direction that the robber crab had really developed an excellent sense to smell in air during its relatively short time as a terrestrial animal. This fact of course prompted us to return to the island several times again and to develop new ways to further understand this very interesting animal that seems to have taken the human niche among the local crustaceans: long-lived omnivore at the top of the food chain.

Returning to the island

Since our first visit to the crustacean paradise on Christmas Island, we have participated in four more research expeditions that dug even deeper into the smell life of robber crabs. In addition, an excellent collaboration with local ecologists Michelle Drew and Michael Smith has allowed long-term experiments. We have continued to track the robber crabs via satellite, but we have also microchipped some of them, like many dog and cat owners do on their pets.

These injected chips have allowed us to get a first real grip on the age of the crabs as we could measure how single individuals grow. They actually grow their whole life, so by measuring how much they grow per year it's possible to calculate their age. The robber crabs can indeed reach the ripe old age of 100 years – and many do!

In later expeditions, we also set up a field laboratory at Dolly Beach, where old males, sixty years and counting, are known to hang out. The beach is full of coconut palms and the males engage in opening the nuts. It takes three days to open one, so once opened it becomes a real prize. Of course, a lot of fighting and stealing goes on in the process, so we wanted to look at how the crabs are attracted to the coconut right after opening.

We opened ripe coconuts with a machete and placed them on the beach. It only took seconds before several big males started to approach from the rainforest. A fight quickly broke out. It was very clear that it was the smell of the coconuts that attracted the crabs, as we could achieve the same attraction with hidden nuts or just the coconut water.

Being olfaction scientists, we of course wanted to know which odours were responsible for both the attraction to the Arenga palm and to the coconuts. So, we brought back samples to our laboratory in Jena, and analysed the smells. At our next visit we could bring the synthetic odours with us and could indeed show that especially one component, acetoin, attracted the robber crabs in the same way as the natural resources. Surprisingly, this odour is contained in both the Arenga berries and the coconut.[9]

The red crabs

It is impossible to describe the animal life of Christmas Island without spending some time on the red crab. The terrestrial red crab looks a lot more like we would expect a crab to look, with a back shield measuring up to ten centimetres. It spends its life mainly in the undergrowth of the rainforest, where most of the soil is actually crab excrements. The red crabs are more or less filling the niche that worms and insects do in our environments: they are detritivores.[10]

When we first arrived at Christmas Island, we came in the middle of the red crab migration. This means that most of the island's more than 100 million red crabs are on the move, swarming across the island, filling up roads, gardens and restaurants. Driving is forbidden to a large extent during this period, except for some roads where specific crab fences and crab crossings are installed. If you do drive, you have to have a person running in front of the car, sweeping the crabs away.

The red crabs all migrate from the mountain rainforest to the beach to reproduce. First, the males move down and establish small territories. The female wave follows, and mate selection occurs. The couples mate and the males return to the forest. The females stay on the beach and wait for the correct time, again very much dependent on the moon and the tide, to drop their eggs in the ocean. Also for this species, the eggs and the hatching larvae need a few weeks in the ocean before they re-emerge as a red, velvet carpet of tiny red crabs moving seemingly in synch up into the rainforest.

Both the adult migration and the returning tidal wave of red crab larvae are absolutely amazing wonders of nature. Other species also make use of this abundance of creatures. After the females have dropped their eggs in the ocean, one of the world's largest congregations of whale sharks circle the island, feasting on all the tiny larvae. On land, robber crabs and nippers take their share of the never-ending supply.

Different crabs – different brains

Crustaceans have very likely entered land in five different events represented by five different types of "crabs". Here, we have looked at the hermit crabs and the "normal" crabs. Beyond these, the crayfish, the amphipods and the woodlice also made their independent entries onto land. All five lineages somehow had to adapt their senses to life on land. Having had our interest piqued by the amazing smell brain of the robber crab, we went on to collect brains from all the other types of crustaceans as well.

To our surprise, it turned out that they have taken very different evolutionary paths in the development of their senses of smell after adapting to living in air. The crabs, the crayfish and the amphipods look very much the same when comparing water-living and land-living relatives. The hermit crabs have greatly expanded brain regions dealing with olfactory input. Woodlice, on the other hand, seem to have dropped the sense of smell completely. Where the primary olfactory part of the brain can be found in water-living species, there is nothing in the land-living ones.

Interestingly, a desert-living woodlouse seems to have "reinvented" the sense of smell, but in a different location.

Why? It evolved an intricate pheromone-based social communication system. Can't have that without a nose. But what came first: the chicken or the egg, the egg or the chicken...[11]

When looking at the five different crustacean types that went onto land you cannot help wondering also why one type expanded its olfactory system immensely, while another dropped it totally. Much more research is needed before we start understanding that one.

The threats to paradise

Christmas Island and its unique ecology are facing two main threats. For about ten years an extremely aggressive ant species, the yellow crazy ant, has been spreading over the islands of the Indian Ocean and the Pacific. These ants form super colonies covering several hectares and eat anything that comes in their way. For the crab communities on Christmas Island this spells disaster. Even though the ants are small, and the crabs are big, the ants swarm over the crabs, spraying acid into their eyes, blinding them in the process.

Over large areas, several crab species have disappeared. Their loss has a direct effect on the rainforest, changing its character from a more open forest to a totally closed one. The rangers on Christmas Island seem to be fighting a losing battle. The poison bait they spread for the ants also attracts the robber crabs. So, each one has to be manually removed before the baiting. Quite a task. Only time will tell if the crabs will survive this threat.

Another threat, which is at least as devastating, comes from human activities. Since large parts of Christmas Island consist of ancient coral reefs, the soil is very rich in minerals. Especially in phosphate. Consequently, the island has been mined for this resource. Mining means cutting all trees and stripping off the topsoil by giant bulldozers.

The soil is then transported in big trucks (which run over the crabs – no one sweeping to help here) down to giant conveyer belts. These in turn run the soil directly onto cargo ships that bring it to far-flung places, including Indonesia, where it helps turn their rainforest into palm oil plantations. A true lose-lose situation, where rainforest is devastated at both ends for short-term profit.

For the animals on Christmas Island, where many species are true endemics, the loss of habitat is of course truly a catastrophe. We can only hope that politicians change their perspective before it's too late.

Chapter 12

Can Plants Smell?

Plants obviously give off scents, much to our delight (on the whole). As plants don't have noses, nostrils or nares, does that mean they can't smell, not even themselves? Just as the receptors inside our noses determine what we can smell, there seems to be a mechanism in plants that picks up on certain chemical cues. Without wanting to anthropomorphize plants, we could still legitimately ask whether plants are using such smells to communicate with each other.

The first serious papers published on such a notion came from two separate sources: Ian Baldwin, now my colleague at the Max Planck Institute for Chemical Ecology, together with Jack Schultz; and David Rhoades with Gordon Orians back in 1983.[1] Their studies claimed that airborne cues from leaves that had been damaged in a way that mirrored an insect attack would travel to nearby plants and affect a change in their biochemistry. More importantly, the change helped the neighbours to battle insect attacks on their own leaves. The claims were hailed in the mainstream press as "talking trees" but slammed by many scientists in the field at the time.

Maybe plants are simply talking to themselves? A much overlooked paper from 1995 was one of the first to suggest that there is some kind of internal communication going on in plants, rather than communication between plants. Volatile emissions could work as internal chemical messengers, or hormones. In the study, researchers at the University of Wisconsin identified a volatile receptor (ETR1 for ethylene receptor) for the gaseous hormone ethylene in the weed thale cress (*Arabidopsis thaliana*), one of the most important models in plant science.[2]

Ethylene is a simple gas that is produced by plants and plays an important role in the regulation of growth and development, particular in cellular processes, but also in physiological processes, including seed germination and fruit ripening. As a plant can both produce the gas and pick it up, it would seem logical that there is some kind of monologue going on. The gas thus indeed serves as a volatile hormone. In commercial operations, the gas is used to stimulate a plant's natural ripening process, most notably for bananas and avocadoes.

Today, almost forty years after the first ideas on plants smelling each other, it's indisputable that they have evolved to exploit certain airborne cues, now consistently referred to as volatile organic chemicals (VOCs), to trigger survival tactics in their own species, but also to set off a chain reaction in their vicinity, either from nearby plants or other living creatures.[3 4]

It's debatable whether you could really class this as communication in the traditional sense. But it is beyond doubt that plants emit chemical cues, and that certain of these emissions can elicit a reaction or response in other plants and living creatures, as well as in other parts of the same plant. Both abiotic stress (caused by environmental conditions including fluctuations in temperature or drought) and biotic stress (caused by living organisms such as fungi and insects) can trigger the signals. Is this interaction a kind of communication in the natural world? What's going on?

Expressed in genes

For odour-released mechanisms to have a chance at working, surely it must be encoded in a plant's genetic machinery. Researchers at the University of Tokyo have recently confirmed that this is in fact the case.[5] In studies on tobacco plants that stretched over a period of eighteen years, the scientists showed that odours activate certain survival strategies. Skipping over the technicalities and going straight to the conclusions, we can summarize that the scientists determined that VOCs influence gene expression in plants.

A gene is expressed when its genetic code is used to direct a set of reactions that synthesize a protein or some other functional molecule within a cell. Copying a segment of DNA to the messenger RNA is the first step in this sequence. This is the gene-reading process, or transcription.

In plants, VOCs seem to influence this sequence of events by attaching to transcriptional co-repressors – the proteins that can turn genes on or off – during transcription, thereby triggering a desired change in the plant's behaviour. Most importantly, the study determined that the odour molecules have to be absorbed into the plant cell in order to cause such a chain of reactions. It would seem then that plants smell directly with the help of their genes, which is a totally new way of understanding smelling overall.

So, this is how the "smelling" occurs. What kind of reactions can this process trigger?

On dire warnings and priming

Could the VOCs warn nearby plants of danger? As the study on tobacco plants also showed, the scent could be perceived as a neighbourly warning. Researchers found measurable increases in resistance to herbivore damage in plants that are growing next to herbivore-infested plants. It's not entirely clear how or why this works, but scientists believe these cues, known as herbivore-induced plant volatiles (HIPVs), play a central role.

While we don't yet understand how plants trigger such warning signals as a defence mechanism, we do know that "priming" plays an important role here. When one plant is attacked it can prime any conspecific neighbours to turn on their own individual defence system proactively. Which generally means that they should start producing chemicals that are poisonous, noxious or repulsive to the attacker, such as a phenolic or tannin compound. Is this a kind of altruistic plant-to-plant messaging in their interconnected world? Or are the plants teaming up to launch a collective defence against a common threat?

While some might eloquently and entertainingly argue that this is the case, including Peter Wohlleben, the German professional forester, in his fascinating book *The Hidden Life of Trees*, other scientists dispute that there is any real communication going on. They assert that this is just a form of eavesdropping amongst plants, leading the listeners to switch on their defence mode when they pick up on any warning signals. As plants are sedentary living organisms, "listening in" and "tuning in" to as many warning odours as possible provides them with a better chance of survival.

Whatever the underlying reason, it seems quite obvious that it's good to be prepared when danger is approaching. Much more research is needed to show if it's real communication that's going on or if it's more a passive spread of information. On the whole, studies generally suggest that plants tend to eavesdrop more on their neighbours than give off signals to warn them of dangers. The reasons for this may be logical – their neighbours are also competing for nutrition and survival in the same neighbourhood, so it wouldn't necessarily be a bad thing if they didn't help them gain an advantage over their own chances of survival.

A sidebar on fungi

Sticking to forests just for a moment, it's also worth highlighting the work of Suzanne Simard, professor of forest ecology at the University of British Columbia, even if it goes slightly off topic. An engaging speaker,[6] Simard has studied the forests of Canada for over 30 years. In numerous long-term experiments, she has demonstrated that trees organize themselves into networks that cover great distances, exploiting the underground web of arbuscular-mycorrhizal fungi (AMF) to survive. This type of fungi hooks up trees and helps them out (and also benefits from being in the network).

AMF connect the roots of the trees as a way to exchange information and resources. This "wood-wide web", as Simard calls it, helps trees put up a defence against herbivores, but also to collaborate and share carbon, nitrogen, and water to trees in need. Simard compares this eco-network and connections in a forest to our human networks, such as our airport system or transportation networks. And, much like our metro systems, much of the networking goes on underground. As yet, there doesn't seem to be many insights on the involvement of volatiles in this network, but this may come.

These fungal systems also help plants and trees to boost their own immune systems. They can help trigger the production of defence-related chemicals. These chemicals then make later immune system responses quicker and more efficient, which brings us back on topic again to "priming".

The ideal defence

Priming leads to fast action, or a response that activates the initial stages in the plant's defence mechanism. This is important, as it will increase the plant's ability to defend itself against herbivores and parasites. It's a kind of trigger to alert a plant's own immune system to get ready. Once a plant has been primed, it will be in a position to defend itself more rapidly and effectively against any similar threats in the future. The triggers for such priming can be live organisms, but also chemical cues, from plant hormones, or VOCs.

What's crucial to the survival of a plant is that this priming can occur at any stage in its development, and in any part of a plant. Priming is basically the state that puts the plant on high alert of an attack by a species of herbivore or a pathogen. Its goal is to prepare the plant to respond appropriately, which means the expression of genes that induce plant defence responses and thereby a greater resistance to insects and pathogens. Can priming be triggered by scent?

Studies have shown that it can. One such study – this time a collaboration between scientists in Sweden, Bosnia and Herzegovina, Italy and the US[7] – suggests that plants can give off these distress scents even when touched, and that this alone will trigger what the researchers call "a rapid defense synchronization" among nearby plants. The defence in this study essentially rendered the plant a less desirable host to aphids – those sap-sucking pests that drain the life out of leaves and stems – even before the enemies can make their approach.

What's more intriguing is perhaps the importance of such signalling on the plant emitting the scent in the first place. Given the fact that odours dissipate in the atmosphere and disintegrate over distances, perhaps the real target of the odour is not the neighbour, but the plant's own bits and pieces. Could the odours work therefore to the plant's own advantage, informing a distance leaf or bud of a danger? Particularly in the case where certain parts of a plant are not well connected by its vascular system, VOCs may likely act as a rapid external signal for inter-plant signalling,[8] again a gaseous hormone.

While this all may be seen as plant-to-plant and inter-plant interaction, could such scents also work with other living creatures?

Tritrophic interaction: a call for help

When they are under attack, plants release VOCs that act as a kind of distress signal. As they are literally rooted to the ground, unable to take flight, this kind of biochemical message is a key aspect in a plant's defence arsenal. For it to work, plants inhabit what could be called an eco-network. Clearly, the distress signal needs to reach its target audience within this network for the plant to benefit. And the target must also benefit from responding to the SOS call. Why go to the bother otherwise?

This concept is called a tritrophic interaction, whereby three organisms are involved in the interaction. Essentially, this means that plants under attack by an herbivore can release a chemical cue that summons help, in the shape of other living creatures – often parasitic wasps – in their eco-network. These creatures come to their rescue, while still benefitting themselves from the interaction by finding tasty larvae to parasitize. Not surprisingly, because of their protective, life-saving actions, these helpers are sometimes referred to as bodyguards.

Each network, target and benefit will vary considerably from the next, depending on the local environment, the species of plant and the species of the herbivore. In general, however, it seems that this type of capability is shared by many plant species. Plants make use of this indirect defence mechanism to protect themselves, and to attract the natural enemies of the attacking herbivores to wound, kill or repel them. This is an efficient way to mount a defence.

Take grass, for instance. Researchers in Texas A&M AgriLife Research[9] have determined that this seemingly basic plant can emit an odour in the form of green-leaf volatiles that create not only that instantly recognizable fresh-cut grass smell to us humans, but also a distress signal to parasitic wasps. Grass emits it when its tissues are damaged – which could be caused by your very own lawn mower, but also by a caterpillar sinking its teeth-like mandibles into a luscious leaf of grass. Green-leaf volatiles are formed from fatty acids and are involved in the complex signalling mechanism in plants in general, working together with the hormone jasmonic acid to prime a plant's defences. This is a fast-response weapon as the scent is released immediately.

In the case of the wasps and the grass, the insects are drawn to the plant by the wafting green-leaf volatiles. They seem to know that the odour means the plant is damaged. And, also that this damage is caused by something they might be in a position to exploit: caterpillars. When they spot the culprit caterpillar, the wasps will sting it, and lay their eggs inside – in the process essentially killing the caterpillar's reproductive cycle and helping out the grass enormously.

It's not just grass that has a defence mechanism based on volatiles. In a collaborative effort led by my colleagues

Jonathan Gershenzon and Sybille B. Unsicker, we determined that black poplar trees emit a particular odour blend when under attack from herbivores, in this case to call on the specific parasitoid *Glyptapanteles liparidis* to attack the herbivore in question, the gypsy moth *Lymantria dispar*.

With a combination of laboratory work involving electrophysiological and behavioural experiments with the wasps, carried out in my laboratory, and field trials in wild poplar forests along the Oder River, east of Berlin, we were able to show that female wasps in particular were more attracted to gypsy moth-damaged leaves than to adjacent non-damaged ones. And that the odour blend released was triggered by herbivory and was also responsible for attracting the wasps.

Interestingly, it appears that the attraction was primarily due to nitrogenous compounds in the blend, rather than terpenes or green-leaf volatiles (although also present). The differences in volatile emission profiles between damaged and undamaged foliage appear to be regulated by jasmonate signalling and the local activation of volatile biosynthesis. The volatiles were again essential cues helping the wasps to find their caterpillar hosts.[10]

In both the trees and the grass, the wasp and the plants clearly benefit from the volatiles. Unlike the caterpillar.

Adaptable under attack

Depending on the plant and the type of attack, plants appear to use allomones, or possibly synomones, to evade or entrap herbivores. Depending on the species of the attacker, plants can adapt their odour emissions to achieve the desired outcome. How so?

Plants don't spit, but they do recognize spit when they've been spat on. When insects tuck in to a plant, they leave traces of spittle along their culinary journey. They do this with a particular purpose in mind. Saliva is used to suppress the plant's defence mechanism, to buy them more time for their meal. However, when the plant picks up on the saliva, it registers the danger posed by the insect and its spit and initiates an appropriate response. (The plant can also be fooled into inactivity by the spittle, as it can also reduce the plant's defence response.) What that response is depends on the plant and the insect. But it will often involve odour messages and that priming we touched on above.

Hostile behaviour

Another way that odour information can be used is in more hostile attacks. We know very well that insects and other animals use plant odours as cues to find their host and food plants. Do we see something similar between plants? Some fifteen years ago, Consuelo de Moraes and her group at Penn State University performed a really interesting experiment with the parasitic dodder plant.

This plant has no photosynthesis of its own and is therefore totally dependent on finding a host where it can plug in and extract energy. The plant produces really small seeds so it needs an efficient system to locate a host once the seed germinates and it cannot take to its wings or feet like other interested parties. So, how to find your host if you're a tiny, sessile seedling growing out of the soil?

Consuelo had observed that the dodder made strange swinging motions when growing out of the soil, almost like it was looking for something. In the first experiments dodder seedlings were put adjacent to tomato plants (which they love), and it was clear that they found their way. When the ability was tested statistically in an arena the plant did very well and grew towards the host in about eighty per cent of the cases. But what kind of information did it use?

The scientists tested several different possibilities and, in the end, only chemical information remained. They designed an experiment where only the smell of the tomato reached the dodder seedling and could then see how it again tended to grow towards the source of the smell. Wheat is not a good host plant for dodder and did not elicit at all the same attraction as tomato.

Which odours might underlie the attraction of the dodder to tomato? The smell of the tomato was collected and analysed, and eight compounds were identified. When the mix was tested, the dodder showed a nice growth towards the synthetic tomato smell. In the end three odours were shown to be the key attractants. In wheat, one odour instead stopped the dodder from approaching with its shoot.

Here then we have a plant that is smelling its way to attack its prey. Watching the movies of the swirling shoot going for the poor tomato reminds me of *The Day of the Triffids*, the 1951 movie where plants take over the world. Here you can really talk about plant behaviour and I think we will learn much more about such active processes in plants in the coming years.

Feeding the world

As our climate changes, food scarcity is becoming a major challenge for humanity. To meet the food needs of a growing global population, we need crops that are more resilient, as well as more nutritious. Could insight into "plant communication" help feed the world in a more sustainable way?

This knowledge of the eavesdropping capability of plants has led scientists to pursue interesting avenues in pairing up companion plants to boost resistance, and consequently production, in certain plant species.

One such sustainable option looked at enhanced pest management using mint plants, which are known to be extremely aromatic even when not mounting a defence to an attack.

Scientists at Tokyo University demonstrated that soybean plants would respond to VOCs given off by neighbouring candy mint and peppermint. The aroma would boost, or prime, the defences in the soybean, and, as a result, lead to less damage by herbivores. This research focused on the diamondback moth, or *P. xylostella*.[11] Being within sniffing distance of the mint primed the soybean to prepare for battle – even when there were no enemies in sight.

In our drive to boost production, is there also a danger that we could end up disrupting the very mechanism in plants that provide our crops with their natural defence?

A devastating loss?

Let's focus for a moment on one of the most remarkable crops in the world: maize (*Zea mays*). In production quantities around the world, it currently surpasses rice and wheat. Crucial for animal feed, and also for biofuels, it is the most important staple crop in many countries.

To meet our need for an ever-more resilient maize production, we have turned to crop breeding and domestication. Along with giving the harvest a boost, could such intense breeding measures also distort the natural indirect defence mechanism in maize: the emissions of VOCs?

This is a complex area and there does not appear to be a quick definitive answer.

On the one hand, there is increasing evidence that crop domestication can alter interactions between plants, herbivores and their natural enemies. Studies have shown that crop breeding could reduce chemical resistance against herbivorous insects. The herbivores get to eat more and go unpunished. On the other hand, such domestication has created more resilient, productive and nutritious plants. Does the upside outweigh the downside?

Studies on this topic often involve comparisons between domesticated maize hybrids and less manipulated land races. In a study of how the maize emits volatiles after attack by a leafhopper that spreads the serious disease corn stunt spiroplasma, scientists could show that the land races emitted many more volatiles – and that they attracted parasitoid wasps very efficiently, a feature that was lost in the hybrids.[12]

Other studies have looked at how to exploit our knowledge of plant communication to help build a more sustainable resistance in the highly productive strains of maize used in industrial farming. In one,[13] maize was treated with cis-Jasmone (CJ), a known volatile organic compound that activates the defence mechanism in plants to produce herbivore repellents. In this study, the researchers investigated whether CJ could prime defence in maize, *Zea mays*, against another leafhopper, *Cicadulina storeyi*, an important vector of the maize streak virus (MSV).

The study showed that the insects preferred the plants that had not been treated with CJ in advance. The treated plants were primed to emit VOCs faster and could thus maybe also attract natural enemies of the leafhopper more efficiently. This could be an interesting development in the story of maize breeding.

And it's not just above ground where such insight might be of use. Studies have shown that there is a lot of action going on underground when it comes to maize and VOCs. In a study that my colleague Jonathan Gershenzon was again involved in, for the first time, researchers were able to identify VOCs that work from the roots of plants.[14]

Maize roots release volatile compounds, in this case (*E*)-β-caryophyllene, in response to feeding by larvae of the beetle *Diabrotica virgifera*, a maize pest that is currently invading Europe. Field trials showed that the odour attracts an entomopathogenic nematode, those soft-bodied beneficial roundworms that will proceed to infect and kill the pest. Interestingly, and unexpectedly, most cultivated maize, notably in the US, has lost the ability to produce these VOCs.[15] Breeding them back may produce more resistant cultivars, Gershenzon notes.

When breeding for increased production in each cob it is therefore important to remember also other factors in the biology of the plant. By maximizing theoretical yield based on laboratory and greenhouse breeding without considering the natural context plant breeders might lose out in the field and less yield might ultimately end up in the farmers silo.

Just a glimpse

This chapter can only offer a glimpse at how plants have evolved to exploit odours to survive in fluctuating environments and networks. Understanding how plants regulate their own defences in nature is an important tool in sustainable agriculture and our economies. Such knowledge could eventually lead to more effective plant conservation measures and more efficient eco-friendlier crop development. In Chapter 14, you can read up on success stories in this area, including a couple of my personal favourites. But first, in the next chapter, you can learn how living things use odours for deception, including how plants use scents to trigger pollination.

Chapter 13

Smelly cheaters

Some sensations make you behave in a certain way with a very high probability. If, for instance, you put your hand on a hot plate by mistake, you will invariably pull your hand back instantly and automatically – without having to give your reaction a single thought. It's a reflex. Smells can cause similar reactions. Some are simply irresistible; they pull you in. While others are just repugnant; they push you away. These kinds of smells are often signals for something that you just cannot afford to miss or ignore.

In humans, such phenomena are rare, if indeed they exist at all. In other species, particularly insects, they are more common. They often work by building on existing pathways or circuits in the brain that, when activated, always trigger a certain predictable behaviour. We call these pathways ecologically labelled lines. The closest we humans might get to such pathways would be our reaction to the smell of food when we're hungry or to the smell of gas in your home. Although both situations will prompt quite distinct behaviours, our reactions are nevertheless still modified by many other processes in the brain. This is often not the case with animals.

In animals with lower cognitive capabilities, these labelled lines have evolved as safeguards. They ensure that an appropriate behaviour is displayed, especially in a life-or-death situation or when a rare resource is at stake. In the chapters on flies and moths I described how these kinds of lines for smell information allow innate behavioural responses in the context of sex, food and enemies.

There are great benefits from having a system that ensures that the correct behaviour is performed more or less as a reflex in a certain type of critical situation. But there is a downside. Having such a predictable response also means that other organisms can exploit it. They can make you do exactly what they want by smelling in a certain way. They might change their smell to fit their own specific purposes. They give off a certain smell too as a means of manipulation.

No rewards

A number of flowers are among the most cunning of the smelly cheaters. Through their deceptive pollination systems, they exploit insects solely for their own returns. Back in the late 1700s Christian Konrad Sprengel, the German naturalist, had already discovered that orchids were cheating insects, by alluring creatures to pollinate them without any reciprocity or rewards.[1] Charles Darwin followed up on this research into unrewarded pollinations in the mid 1800s.[2 3 4] I have also worked on a number of these systems. Each time I have been just as astounded as Sprengel and Darwin by the deviousness of the deceit.

In the end, every time the deceit comes down to exploiting one of three specific smell pathways in the insect brain. These are the pathways that trigger feelings of attraction in the insects based on either the smell of food, the odour of egg-laying sites or, the most alluring scent of all, sex. Coming back to our definition of semiochemicals, this kind of odours are clear examples of allomones, benefitting the sender.

Smart deception

Certain kinds of orchids have evolved to smell (and look) like the perfect female. When we studied these flowers, we found that their smell exactly mimics the sex pheromone of *Andrena* bees. The pheromone produced by the bees contains multiple components. The flower copies these exact same compounds in exactly the right proportion.[5] But the cheating doesn't end there.

The flower has also evolved physically in such a way so that it visually resembles a female bee, thereby further increasing its overall resemblance. The similarity is so convincing that the male bee will try to mate with the flower. In the process, as the bee tries to latch on in the mating position, orchid pollen gets stuck to his body. This pollen will be transferred to the next flower when he finally realizes he's not getting anywhere.[6]

Now, bees are very smart insects. So, the orchids have to be smarter still. Especially as the female of the specific bee species that pollinates the orchid only actually mates a single time. This means that the male is determined not to waste any time by trying to court females that are unreceptive. How does he avoid the chances of unsuccessful mating? He has evolved the capability to learn and distinguish the minute differences between the smell of individual females. Thanks to this skill, he never tries to mate with the same female twice.

If every flower of the orchid smelled exactly the same, surely the deception wouldn't work twice. The male bee would not attempt to copulate with the deceitful flower a second time, and, as a consequence, would not transfer any pollen from the first to another. But he does! The question is why?

When we analysed the flower odours, we found exactly the same variation between the flowers as is found between the female bees. The male bee therefore thinks each flower is a new, unmated female. He is fooled over and over again, each time faithfully transferring the orchid pollen without getting any reward whatsoever. He just wastes valuable time and energy.[7]

CHAPTER 13

The smell of death

Orchids aren't the only cheats in the plant world. Among the *Arum* lilies, other types of deception have also evolved. Our venture into this research area also shows how serendipity can work for science.

In 1989, I was involved in setting up a conference that would specifically deal with the sense of smell and taste in insects. The conference often takes place on the island of Sardinia. Over the years, we've established a popular tradition. We often take a boat excursion out to a couple of small islands off the coast. On the boat, we'll enjoy ample measures of seafood and wine, to get us into the right mood to dare swim the 200 metres ashore.

One time, when we subsequently explored the little island, we encountered some very strange flowers, both when it came to looks and to smell. The visual impression was of a giant, meat-coloured calla lily flower, but the smell was so repugnant that it reminded us of a rotting corpse. As chemical ecologists and smell scientists, we, of course, had to investigate the reason for this unusual floral bouquet. This was how the project on the dead horse arum got started.

After collecting the smell of the flower both on site, on the islands off Sardinia, and in the laboratory, we set out to identify the compounds that were produced. Then, by linking chemical instruments to the antennae of flies we had found buzzing around the flowers we could see which compounds were indeed smelled by the flies. We then collected and tested the smell of rotting pig meat. We found it had exactly the same compounds as the dead horse arum. The flower was indeed mimicking a rotting corpse.[8]

So, what was the ecological background to this mimicry, and why would a flower smell, to our nose, so badly? The flies buzzing around the big flowers were flesh flies. The female of these insects is totally dependent on finding rotting meat to lay her eggs. Consequently, she has developed an ecologically labelled line to detect and respond to such odours. The smell of rotting meat is for her a super-attractant. We next tested the odours in the field and could show that female flies were extremely attracted to the smelly stuff.

The source of pollen in this flower was hard to access. The flies had to enter into a trap chamber from which a kind of tail was protruding. To test the function of the flower we made copies and perfumed these with the rotten odours. As expected, the flies came and landed on the surface of the flower. But they never really entered into the trap. Something was missing. Some other factor had to be present that made the fly enter inside the chamber. What we discovered was astonishing.

At the entrance of the trap chamber it not only smelled like a rotting corpse, it felt like one. The tail sticking out of the trap chamber was warm – 37 degrees centigrade warm. When we simulated these exact same conditions together with the stink, the flies faithfully entered inside. Down in the chamber, the flower trapped them there for some time, ensuring that before they could escape they would be covered with pollen. This meant that they would then move to the next flower to inadvertently complete the pollination.[9]

Evidently, the flower has evolved every means possible to perfect the imitation of a rotting body. It gives off the correct odours, it looks exactly like the meat you would find at the surface of a hairy animal, it is nice and warm, and feels just like the heat produced during the breakdown process in a dead body. By mimicking all the key stimuli, a female flesh fly is looking for when searching for a place to lay her eggs, the flower gets pollinated for free. The fly is the big loser, as, much like the bee in the story with the orchid, it again wastes precious time and energy, sometimes even its eggs. If the flesh fly falls completely for the con, it will lay its eggs in the chamber of the plant, a place where its offspring will surely die.

You would think that the fly could evolve to escape being exploited in this manner. Why doesn't it develop a more acute sense for what is real and what is not? The cheater system very likely persists through evolutionary time thanks to several different factors. The first has already been mentioned: it mimics a signal that is too important to ever ignore. Secondly, the plant only flowers during a limited period per year on a few small islands in the Mediterranean. This means that the flower puts pressure on the flies only during a small window in time and space. The rest of the year, and in the rest of the world, the flies will find what they expected, a dead animal. The pressure to change or evolve is therefore kept to a minimal in the fly.

Smells nourishing

Another arum lily, the Palestine arum, which is found in the driest areas in the Levant, helped us to understand the key compounds that attract vinegar flies to wine and balsamic vinegar. Whenever you look inside this kind of lily, you'll always find a large number of fruit flies. Simply sitting there. Just like the flesh flies in the dead horse arum, they are also carriers of pollen. It turns out that the Palestine Arum produces exactly the key compounds found in wine and balsamic vinegar. In this way, it mimics the perfect food for the little flies, lures them into the flower, and again gets them to perform unrewarded services to the fly's own detriment.[10]

Interestingly, in another population of the same species we did not smell the wine-like odours at all. Instead, the flowers smelled of horse manure. When we investigated which flies were attracted to this particular flower, we found that big manure flies were indeed the dupes here. We had stumbled upon a case of ongoing speciation. As the flowers smell differently, they will never cross-fertilize. The flies will only fly to flowers – and subsequently transfer the pollen to them – that smell the same. Over time, this will very likely result in two new species of the Palestine Arum instead of one.

Another food-based deception occurs in an *Epipactis* orchid. This flower specializes in attracting hover flies. Their larvae live from eating aphids on plants. To find the minute sap suckers, the female hover fly has developed a keen sense of smell for their pheromone.

Where there are aphids, there will be the smell of their sex attractant. In turn, the orchid has evolved an odour exactly mimicking the pheromone of the aphid. In addition, small red-coloured bumps inside the flower look just like aphids. Again, the female is attracted in by an irresistible odour that signifies a very important resource: food for your kids.[11]

What these examples have in common is that they show evolution in action. They confirm that different flowers have evolved the capability to make use of innate behaviours in insects that are triggered by specific sensory stimuli – especially by smells. Their capability can target sex behaviour, food attraction or the search for perfect egg-laying sites. All are vital resources for survival and reproduction and therefore more or less impossible to ignore.

Dangerous liaisons

Predators are also known to exploit attractive smells to catch their prey. The bola spiders are masters of deceit in this context. They make use of the male moth's absolute fixation on sex once he has got wind of the perfume of a female (Chapter 7). The bola spider has developed a very special hunting method. It does not weave a web like many other spiders. Instead, it constructs a sticky ball out of its silk. This ball hangs in one long thread and dangles below the spider. On this little ball, the spider deposits exactly the correct pheromone of a moth species. It then sits and patiently waits, with its fishing line dangling in the wind.

When a duped male moth approaches close enough the spider swings its bola and hits the moth, which gets stuck on the ball and can be hauled in and devoured. There's also evidence that one and the same spider can change the chemical composition of the pheromone mimics applied to its bola over the season. That way the pheromone will always perfectly fit the occurrence of different moth species in its surroundings. But, in fact, the pheromone has now turned into an allomone. Interspecific messenger benefitting the sender.[12]

CHAPTER 13

Sickly attraction

Microbes and unicellular animals are smart at deploying insidious strategies based on smell. By affecting the odour production in their hosts, for instance, they can make these more or less attractive to each other, or to other organisms. When I work on insect smell ecology, I always try to think like an insect – to put myself in their context. This is the best way to find new mechanisms based on smell.

In one case, I was pondering the idea that to escape contagion, you need to avoid sick friends. If you sit beside a friend who is sneezing and coughing, you might want to move a few seats away to dodge the virus particles. Our hypothesis was that the same might hold for vinegar flies. To test this, we first infected flies with some bad pathogens and observed what happened. The result was the complete opposite to what we had expected.

Healthy flies were strongly attracted to their sick friends and to their faeces. They also loved food spiced with faecal pellets from infected flies. The females laid their eggs where sick flies had been before. Their behaviour meant that the healthy flies got infected and died. The same held true for the larvae coming out of the eggs that had been laid on the sick fly substrate. This was of course a big surprise. Why would you voluntarily run the risk to get infected by a deadly germ?

As we're smell scientists, we naturally started looking at the odours of the healthy and the sick flies. Healthy flies emit low amounts of a very attractive sex pheromone. When we analysed the smell of the sick flies, we found that they produced 20–30 times more of this pheromone than the healthy ones. As such, they became super-attractive in every way to their friends. We then checked if this really meant that the microbes were spread. This was indeed the case. The flies were evidently strongly attracted to infected flies and were in turn infected themselves. This means that the microbe "hijacked" the pheromone production system of the fly, cranked up the production and ensured its own propagation in the fly population.[13]

In another similar case, scientists have investigated the smells that attract mosquitoes to humans. After collecting the chemicals emitted by different parts of the human body the investigators could identify specific compounds that are used by the mosquito female in her search for a meal of blood. An important component of this bouquet is carbon dioxide. When healthy humans were compared to individuals infected with malaria, the sick people turned out to be significantly more attractive than the healthy ones. It seems that the malaria parasite uses a similar strategy as the fly pathogens to ensure its spread among humans. More about this in Chapter 9.

CHAPTER 13

The human side

We humans also like to make use of the attraction of other species to odours. In Chapter 14, I will look at the applied use of odour signals and discuss how insect pheromones are used in different ways. To protect, for instance, our harvest. But I will also show how we use smell attraction in more direct ways.

When I go hunting for wild boars in the Swedish forest, I often put out some bait to lure the animals out. Some corn or sugar beets are very appreciated by the boars. This food-based deception can, however, be strongly intensified by applying beech tar on a stone or a tree close to the baiting site. The smoky smell of the tar is quite irresistible to the boars. They do their best to smear it all over themselves by rubbing against the stone or tree. We can only speculate as to why they do this, but it could be linked to the fact that many animals like to camouflage their own odour. Another reason might be that the strong smell repels skin parasites and bloodsuckers. In any event, their attraction to beech tar helps put dinner on my table.

You'll find similar strategies at use when fishing for sharks. If you pour some blood in the water around the boat, the smell is thought to attract the sharks from afar and bring them close to our baited hooks.

All of these cases show how very strong, hardwired odour signals are open to exploitation by other organisms.

This can happen to obtain a service from the dupes, such as for pollination, but it can also be to attract prey to kill, as with the bola spider and myself. One thing that all of these cases signify is the importance of odour information for many, many organisms.

Often an instant reaction to a smell can be the difference between life or death, between spreading your genes or not – or to being taken for a fool or not.

Chapter 14

Exploiting Smell and Smelling for Our Own Purposes

Smelling and smells are used to gain information and to influence the behaviour of others in many, many ways by us humans. We collect information by our own nose, by other animals' noses and by machines. We influence the behaviour of our fellow humans, but also that of other animals and maybe also of plants. It's good to be aware of all these possibilities, as you are probably exposed to them on a daily basis.

Collecting information

The most obvious way to collect information about our chemical surroundings is, of course, through our own nose. It's where we carry one of the most exquisite instruments for chemical analysis that exists. I was just reminded of this recently, when a friend who is a connoisseur of French wines brought two very nice, identical bottles to go with a dinner of game that I had cooked. She opened up the first bottle and poured me a glass.

Even before I had taken the first sip, I knew that this bottle could go directly down the sink. The unmistakeable unpleasant smell of geosmin, a.k.a. corked wine, hit my nose, and I couldn't hide my negative reaction to the odour. Think of wet dog and you will know what I mean. Fortunately for us both, the second bottle was of impeccable quality, so the dinner was saved.

In the same way, we are constantly analysing every food item, every drink, but also every environment where life takes us. Just remember how efficiently we smell gas, smoke, mould, putrefaction and all other warning smells. But, equally, all the good things in life, such as fresh spices, old whisky and your mother's meatballs. Much more is written about these capabilities in Chapter 2, but in the context of this chapter, the focus is on the best smelling device that we carry with us at all times – and just how our sensitivity to smells can be used to manipulate us, too.

Using humans to smell

Some of us go beyond the everyday smelling and become "super noses". Usually, such a skill comes with training, not by virtue of being born with a better sense of smell. Super noses are very important in product development, both in the food and drink industry, but also in the very successful perfume, deodorant and designer-smell industry. Car manufacturers are known to test components and materials in their odour labs to get that unique smell of a new car just right.

All of these industries have either their own panels of smellers, or they hire companies that specialize in these kinds of evaluations. In both scenarios, a group of trained people are exposed to new smells and tastes under development. They are to blame or thank when you have a slightly new experience on opening your favourite ketchup bottle or the door to a new car.

Another aspect of human detection is the smell of disease. Below you will find examples of how animals and machines are used, but the first detectors were probably humans. Nurses would sniff the urine of potential diabetics to smell the diagnostic odours. More recently, one woman claims to be able to detect whether someone has Parkinson's disease just from their musty odour, and this before any symptoms appear or a firm diagnosis is made. Scientists at Manchester University in the UK have been working with the former nurse in their search for an early diagnosis tool. Pinpointing the disease's signature odour and its volatile biomarkers could be an important step in the process.[1]

Using animals to smell

For thousands of years we have made use of animals to help us out when our own sense of smell falls short. This happens especially when it comes to sensitivity, where many other animals have developed a considerably stronger sense. The classical example is the hunting dog (see Chapter 3), where we use the dog's superior nose to detect scent left by prey.

Some species are extreme hunters and will follow deer, foxes, badgers and other animals for hours. Recently, my own dachshund – despite being eleven years old – disappeared into our local swamps and was out hunting some animal for a full four hours. It ended with my wife having to pull him out of a badger's den by his tail. The smell-driven hunting instinct in dogs is evidently strong, but over the centuries we have increased it in our breeding programs, often for our own purposes.

A dog's willingness to pursue and find items by their smell can be used for many other tasks. The classic example is the police dog. A constant partner of our law enforcement squads, dogs were initially used mainly to track people down from crime scenes or after prison escapes. Today, they also smell out drugs, money and dead bodies, even under water. In the military, dogs have been put to use to track down mines in front of advancing troops or to clear out mine fields after the conflict.

In public health, dogs play an increasingly important role. Over the years, our canine friends have been shown to be able to detect early stages of cancer, and even to differentiate between different types of the disease. More recently, the upbeat story of golden retrievers dressed in bio-detector jackets made the media rounds.

They had been trained to identify Covid-19 infected people arriving in Chile. In the UK, researchers from the London School of Hygiene and Tropical Medicine, in collaboration with charity Medical Detection Dogs and the UK's Durham University, are also trialling dogs for sniffing out Covid-19 in public places. They are not the only ones.

A dog can basically be trained to sniff out anything that emits even the faintest smell, as can rodents. We've also put them to human use. Giant sugar cane rats, for instance, have been replacing dogs as landmine detectors. The only problem with this solution is that the training takes about the same amount of time as it would with a dog, but the rats have a much shorter lifespan. And yet, they can achieve a lot in their lifetime, as one famous rat can testify.

In 2020, for the first time ever, a rat received the prestigious PDSA Gold Medal, which is awarded by the British veterinary charity People's Dispensary for Sick Animals (PDSA) for animal bravery (dogs are usually on the receiving end). Magawa, an African giant pouched rat, was chosen for its "life-saving devotion to duty, in the location and clearance of deadly landmines in Cambodia". So far, Magawa has cleared 141,000 square metres of land from 39 landmines and 28 other unexploded ordnance.[2] Not bad for a rat!

Using insects to smell

Can insects be trained to help us in detecting specific smells? Just like dogs and rats, insects can be trained to detect more or less any odour. Again, the smell of explosive (typically TNT or cyclohexanone) is often the target substance. The US Defence Advanced Research Projects Agency, DARPA, has been known to invest heavily in finding ways to train insects to detect landmines and to subsequently land on them to mark their location.[3 4]

A commercial company has even tried to sell such a system for scanning passengers and luggage at airports. Restrained bees in a box can be trained to extend their proboscis, their tongue, when stimulated by a specific odour. This is done by pairing the odour with a reward, such as sugar water. After just a few pairings the bee will be conditioned to extend its tongue in expectation of the sweet reward. The detecting system contains a row of restrained bees that have been trained to detect a certain odour, such as TNT. When they smell the odour, they extend their tongues, which hit a tiny laser beam and set off the alarm. The drawback is of course that bees don't live very long, so the training has to be carried out over and over again. As far as I know, the business venture was not a hit.

Machines that smell

Animals are great smellers and they can be used for many tasks. On the downside, they have to be trained and they have a limited lifespan. It would, of course, be great to have access to smelling machines that work 24/7 without training and for a very long time. Such machines do exist for very specific tasks. Returning to the airport, today you might have to pass through a giant "smeller" that picks up traces of TNT or other telltale molecules. This is just one example of the very many options that are currently available on the market.

Basically, an electronic nose is composed of hardware and software. The hardware has a detecting part with sensors adapted to the type of odours being detected. The sensors can be built with various techniques and materials, including metal oxides, conducting polymers and quartz crystals. Their task is basically to change their properties based on the chemical characteristics of the odour. They are the olfactory receptors in the system. The output from these detectors is fed into a computing part, the brain, where different types of software try to make sense of the input and in the process provide an identification of the odours that hit the detectors. Today, as the machines get more and more sophisticated, machine learning is often involved.[5] [6] [7] [8]

The market for electronic noses is huge. In the food and drink industry, we use machines to routinely check quality; in agriculture and forestry, we need them for quality control too, but also to detect pests and pesticides. In medicine, they help diagnose diseases, such as cancer, infections and tuberculosis. We also see the machines in use to monitor our environment both indoors and outdoors, and, as mentioned before, electronic noses make up an integral part of many of today's security systems.

These artificial smellers still suffer from many problems. The main one is their inability to achieve a sensitivity that in any way matches real systems found in nature. The other is to achieve a system that can detect a high number of molecules that are often chemically unrelated, but highly relevant. Thanks to their very high number of receptors, this is what real noses do.

It's more or less impossible to put such a number of relevant artificial detectors on a machine and then to also make it smart enough to make sense of the input. Several technical aspects of the sensors themselves also pose serious challenges. But, as in all other areas, progress is fast. New, smarter machines are unveiled every year.

Manipulating human behaviour

The sole purpose of one of world's largest industries is to make humans smell non-human. Ever since medieval times, human smell has been considered primitive and a sign of class. Those who could afford it invested in other smells, and a whole guild of perfumers grew up, most notably in France. These professionals were constantly busy, racing to invent new, exciting mixtures, and the competition continues today. Just venture into the tax-free shop at your nearest airport or into your local department store, and you'll be confronted with a seemingly never-ending choice of smells to apply to your body, smells that typically have nothing to do with humans in the first place. Flowery and fruity smells are very popular as main notes in the perfume, but under these some human smells might be hidden in the minor notes. These are sometimes of surprising origin, including faecal and urine smells.

In 2018, the perfume industry was valued at more than US $30 billion, and, depending on which source you consult, it's expected to hit US $50–90 billion in 2025. That figure is just an indication of how much we are willing to invest in disguising our true smell.

Would it be possible to fabricate a perfume that would make you absolutely irresistible to the opposite sex? As we saw in Chapter 2 on humans, the idea of human pheromones is still very controversial. So, the answer is no, it's very unlikely. But this doesn't stop this market from booming. Just google the word pheromone and perfume and you will get a number of offers that promise great success at your next visit to the local bar.

Another way to manipulate our behaviour is to make us more willing consumers of certain goods. This is not a new invention. Bakeries have long since made an effort to direct their exhausts towards the street. Passers-by are reminded of the wonderful bread for sale inside and might enter to spend their money more often.

Today, customer manipulation by smell has become much more advanced. The "real stuff" is sometimes not even around where the goods are sold. Some years ago, I was working in Japan. After indulging myself with superb sushi for some weeks I was ready for a frozen pizza. When I was choosing which kind to bring home, I suddenly picked up the distinct smell of freshly baked pizza hitting my nose. The supermarket had an exhaust for synthetic pizza smell placed right over the freezer! It worked. In a similar manner, the smell of many food and drink items can be synthesized and used to trigger our willingness to invest.

Similarly, smells can be used to transport us to a specific context, or just to create a generally positive atmosphere. Coconut smell in a travel agency foyer, for example, allegedly increased the sales of trips to exotic islands significantly. "Green" smells of grass or trees are thought to improve the working performance in offices. This kind of smell manipulation and olfactory design has grown into a major industry, with an annual turnover in the millions of dollars.

Manipulating animals

We use smells on a daily basis to trap or to change the behaviour of many animals. Just yesterday I returned from my wild boar baiting site, where I smear beech tar onto a big stone. The boars just cannot resist this smell and come and rub themselves against the stone to add the tar odour to themselves. Why they would want to do this is not really known. It could be a way to counteract ectoparasites but also to camouflage their own odour. Anyway, together with some tasty corn cobs the tar smell becomes a very good attractant to bring the elusive boars to where I want them.

But maybe the most obvious examples of smell manipulation of animals come from the insect world. The simplest form being the different types of trap that you can buy in your local hardware store to control ants, moths, flies, cockroaches, and so on. All these products are based on research revealing which odours are indeed attractive to these different types of insects. Mostly, the traps are baited either with synthetic food odours or with synthetic pheromones. Trapping has also been attempted at grander scales to control insects in the field. There, the traps are usually baited with sexual or aggregation pheromones luring the insects inside.

As we saw in Chapter 9, without doubt the biggest insect challenges for humankind are vector-borne diseases, most notably those transmitted by tropical and subtropical mosquitoes, including malaria, dengue fever and the Zika virus. Several management methods are built on our own attractive smell. Bed nets impregnated with insecticide, for instance, are hung over sleeping people in infested areas.

The mosquitoes are attracted by the smell emitted by the human under the net, but land on the net and come into direct contact with a deadly insecticide. Quite simple, but also very efficient.[9]

Another, similar strategy is the so-called eave tube. Mosquitoes tend to enter houses in Africa through the open eaves. For this method to function, all openings of the house are closed by windows, doors and insect nets. Special tubes with an insecticide net are then mounted on the only entrance port through the eaves. The mosquitoes are attracted in by the human smell, again land on the nets in the tubes, and die.[10]

In another large-scale attempt to decrease village populations of malaria mosquitoes, my colleagues Rickard Ignell and his co-workers discovered that these mosquitoes are strongly attracted to cow urine as a source of nitrogen. Consequently, they constructed mosquito killer traps built on such urine odours. In this way, they were able to decrease significantly both the number of mosquitoes and the malaria infection rate.[11]

Several more expensive products are available for mass trapping of mosquitoes. One of the most well-known is the Mosquito Magnet, which works more or less like an attractive vacuum cleaner. Mosquitoes are lured in by a mixture of CO_2 and 1-octen-3-ol, a typical mammal smell (see Chapter 9), and sucked into a container, where they are trapped. This machine has turned out to be quite efficient as home protection when mosquitoes get really annoying.

In several other cases, mass trapping has had limited success, as most insects in an outbreak situation occur in such numbers that our efforts to trap them just don't suffice. A good example involves the spruce bark beetle[12] (see Chapter 10) and a major mass trapping programme was attempted in Sweden during the severe outbreaks in the 1970s. Unluckily, the damage to the forest was still considerable, despite millions and millions of beetles being caught. This year, I'm trapping in my own forest, but I have a feeling that the effect is mainly therapeutic for the forest owners, seeing that, somehow, they feel they're hitting back at the enemy.

More advanced methods to manipulate insect behaviour in agriculture and horticulture have been more successful. A simple way is to use attractants just to let you know when the pest insect you're targeting is around. This means that the use of pesticides can be restricted to the weeks when it's really needed.

Another, more elegant, method is built on mating disruption in different moth species (see Chapter 7), where minute amounts of pure pheromone are emitted from many sources, including in vineyards or apple orchards. This means that the whole environment smells like a female calling for a male. After a while the males just seem to give up and no matings take place. You can compare the situation to a city full of millions of perfect (artificial) copies of women, impossible to tell from the real thing. Among them there might be a few real ones.

From the male perspective you can easily understand that the moth man quickly gets exhausted trying to track down the real thing. A clearly beneficial aspect of mating disruption is that it is allowed in ecological farming, which allows the farmer to sell at a higher price and at another target market. While mentioning pheromone-based management methods we should not forget the lampreys, which we looked at in Chapter 5. The pheromone of these fish has been identified and attempts to use it to lure the female away from commercial fishing areas are underway.

Boosting food production

An even more appealing (but much more work-intensive) method has been developed at the International Centre for Insect Physiology and Ecology (ICIPE) in Nairobi, Kenya. I've had the pleasure to work there since the early 1990s and to sit on the Governing Council for quite some years. Zeyaur Khan and John Pickett have developed a method based purely on natural odours. Let's take it from the beginning.

Today, maize is the main staple food in large parts of Africa. However, it originates in South America. When the crop was introduced to Africa some of the local insects found it to be an excellent new resource. Especially the stem borers. Their larva, as its name implies, eats its way inside the corn stem and causes it to collapse. This habit totally devastated the crops for smallholder farmers all over East Africa.

My colleagues in Kenya and Britain had the bright idea that there might be some local plants that would be more attractive to the stem borers than corn and that there might also be some plants that smelled awfully to the insects. Using their knowledge of moth ecology they identified some good candidates and started testing.[13] After some less successful trials, as always in science, they found two candidates: a local grass that was the natural host for the moth before the corn arrived, and a kind of bean, the smell of which, for some reason, the moth hated.

After many tests, a regime was established where a rim of grass was planted around the cornfield. In between the rows of corn, beans were planted. And it worked incredibly well. The moths were pulled out of the corn by the wonderful smell of the grass and subsequently laid their eggs there. At the same time, they were pushed out by the awful smell of the beans. The Push-Pull technology, or Sukuma-Vuta in Swahili, delivered much more than expected.

The harvest increased manifold, not only by the decrease in moth attacks, but also by the nitrogen fixation by the beans. At the same time, the beans turned out to prevent germination on the seeds of a serious parasitic weed, *Striga*. To add even further to the benefits, when the grass surrounding the crop was cut and fed to the farmers' cows, their milk production doubled. A win-win-win situation. Have a look at www.push-pull.net for more information.

Just a final word on the push-pull story. If you're lucky as a scientist, you get to experience that moment when science has made a real difference to people's lives. For me, the moment came when I was standing in a village close to Lake Victoria. I was talking to an old man, probably younger than me, who explained that thanks to Sukuma-Vuta he could not only feed his family, he could also see some surplus to send his grandchildren to school. And on top of it all, he could help his neighbour (who hadn't adopted the technique) with some corn.

However, a project like Push-Pull requires a long-term scientific perspective and collaboration between many scientific disciplines, from chemists to social scientists.

Battling the flies

Another push-pull strategy aims to control the tsetse fly. These flies are a major problem in large areas of central Africa, as they spread the sleeping sickness both to humans and livestock. Again, my colleagues at ICIPE got help from a well-known savannah fact: everyone hates the smell of the waterbuck! Most predators avoid catching waterbucks and, more importantly, tsetse flies don't bite these animals. The scientists went ahead and collected the smell from waterbucks and could isolate a five-component mixture that really smelled bad to the flies.[14]

By hanging little containers with this mixture around the neck of the local cows, the infection rate of Nagana, cow sleeping sickness, was brought down dramatically and the animals remained healthy. In parallel, a smell that the tsetse fly really loves was formulated. This is the smell of the buffalo and its urine. Specific tsetse traps were constructed luring the flies into contact with an insecticide. The tsetse flies were thus pushed away from the cows and the villages and pulled in to the traps. Another push-pull situation that functions really well.

As seen from the push-pull results, repellents can also be a very efficient way to manipulate insects. We have all added some mosquito repellent to our skin and some might have tried just to rub the arms and legs with a lemon or to put coconut oil into the dog's fur. All of these methods aim to repel annoying insects and ticks. The typical insect repellent contains DEET (Diethyltoluamide), which is an absolutely artificial compound invented by the US military in the 1940s to protect their troops especially during jungle warfare.

Thousands of compounds were tested by sending soldiers with naked legs out into mosquito-infested swamps, and then calculating which compound decreased the number of bites. In this way, the army scientists found that the non-toxic DEET had a drastic effect. Why, was unknown until very recently, when scientists could show that DEET probably functions by blocking the mosquito's smell receptors for human odour. Today, the question regarding the non-toxic DEET is also under debate and prolonged use is discouraged.

A number of essential oils, among them from lemon grass, lemon, mint, catnip and rosemary, also have some effects on mosquito attraction, but none has so far been able to compete with DEET. Another strategy for personal protection is to carry a little gas burner, more or less like a large cigarette lighter, that evaporates Δ-allethrin. These compounds are related to synthetic pyrethroid insecticides and keep mosquitoes away quite efficiently. A number of fatty acids present in coconut oil have also been shown to be active against biting insects and ticks. Pet owners are recommended to rub dogs and cats in this oil to prevent tick and fleabites.

Follow your own nose

Above you read that both dogs and artificial noses are used to smell out different types of diseases. Another way to use smells to diagnose disease is to use the nose of the actual patient. In the Covid-19 pandemic it has been shown that failure to smell and taste is one of the telltale symptoms of infection. A trustworthy method to use this fact in diagnosing the disease is, however, still lacking. This is not the fact for some other diseases. For both Parkinson's and Alzheimer's disease it has been shown that decreased olfactory capabilities is one of the very first signs that shows up in a patient. Building on this knowledge, special sets of odours formulated in easy to use "Sniffin' Sticks" are used to find early signs of disease.[15]

Our nose is the prime tool to judge the quality of odours. It's directly connected to the, so far, most amazing data-processing centre in the world: our brain. Trusting our nose, we can avoid many hazards in life, but we can also find some real treats. Using the nose of our best friend is also an excellent way to boost the sensitivity of detection but then we also have to collaborate closely with our dogs, as we have done for millennia. The development for both these detector systems is probably only limited by our fantasy.

When it comes to artificial smelling, machine learning and artificial intelligence will play a significant role in the future. By using such systems, we can possibly start emulating our brain's ability to combine information from many, many sources (olfactory receptors) to form the Gestalt of a specific odour (Chapter 14).

When it comes to the manipulation side, I'm convinced that the industry for both perfumes and food odours will continue to thrive. We are so hooked on these different smells and will continue to pay top dollar for a few ounces of Chanel No 5. If we learn more about how animals use odour communication, we can also develop new systems that enable us to move away from insecticides, herbicides and all other poisons that we've become so good at spreading around us.

Simply put, we need to keep our eyes (and noses) open to new smells, new applications and new technologies that will help us in our daily life as humans. By learning something about how we and our fellow human beings on this Earth sense and emit smells, we can already find new and simple applications, be it to attract a tiny fruit fly or a massive wild boar.

Conclusion

Smelling the Future

In every chapter of this book I have provided examples of how important smells and smelling are for different organisms, from insects to humans. Smells guide you to food, mates and, if you're an insect, to good places to lay your eggs. But they also warn you of dangers in the form of spoiled food, enemies or fire. In the first chapter we saw how the smellscape has changed during the era of Anthropocene, and I provided some examples of how smells and smelling can be put to use in different ways in the final chapter.

What will happen to smells and smelling in the future? Will human activities continue to change our environment so dramatically that we will see profound changes? Or will the interactions shaped by evolution remain functional and intact? Will our use of smells develop into new, maybe unexpected, directions? Or have unintended consequences? Will we be able to communicate remotely with smells? In this short conclusion, I will dwell a little on possible scenarios for our future smellscape.

Since the beginning of life our living environment has undergone a slow but constant change through evolution.

A process that Richard Dawkins describes as analogous to having a blind watchmaker in control in his book *The Blind Watchmaker: Why the Evidence of Evolution Reveals a Universe without Design*. Through spontaneous mutations, nature is testing and testing and testing new variants. Most of them are worse than the previous version but some provide the carrier with an advantage in survival and/or reproduction and form the material for positive change.

These processes will also affect the natural odours that surround us. Flowers might slowly change their bouquet to achieve even better pollination services from the insects or to better stand out among all the chemical noise created by humans. The same might hold for other communication systems that change in ever ongoing speciation processes, or that again need to compete with the very many other smell molecules floating around in the air.

In our own attempts to manage different types of animals we also put a lot of pressure on the smellscape, including pheromone communication systems. If everything in a field smells like a female, there is a great advantage to be a female that stands out by smelling a bit differently – and to be a male who is attracted to that special female. They will be able to find each other even in the man-made fog of fake aphrodisiacs.

The more dramatic change is of course all the extinctions of plants and animals that human activities are causing. The odour impressions from all these organisms will be lost, just like the animal or the plant itself. In a museum, we can still get a visual impression of the Tasmanian wolf, but the smell of one is very hard to recreate or even have a vague idea about.

The non-living environment might also change. Sudden eruptions of volcanoes can release clouds of sulphuric gases, and who knows what kind of odours extraterrestrial material falling onto our planet might carry. The ocean is currently undergoing dramatic changes due to human activities and might also change its chemical emissions in the future, providing a new oceanic bouquet. Water in itself doesn't smell, but all the things living in it do. If our constant pollution of the world leads to a shift in the balance between microbes, we will witness changes to plankton, seaweed, fish, and so on, and thereby also to the smells. Just think of the dimethyl sulfide that we discussed in Chapter 5. If this emission changes because of a shift in the phytoplankton flora, it will have many consequences.

All of these changes are underway now and can become even more dramatic in the future. The smellscape can indeed become a totally different one, but very likely at quite a slow pace.

But what about our own use of smells? Both for manipulation and communication there might be some interesting changes coming up in the future. In Chapter 2, we looked at how humans smell and how we might possibly communicate with each other using odour messages. For quite some time now, other kinds of messages, such as language and gestures, can be transferred quite faithfully by technology. Alexander Graham Bell made the first phone call in 1876 and John Logie Baird transmitted the first television in the mid 1920s.

And yet we are still unable to transmit even simple odour impressions, not to mention complex and dynamic mixtures. When you're having a video call with someone you love, you transmit how you feel both with words and with facial expressions, but the pheromonal information is definitely missing. How could such a pheromonal transmission be imagined?

Well, first you would need a machine sniffing the odour you're emitting, much like the microphone for sound or the camera for visual impressions. As I described in Chapter 14, a host of electronic noses, built on different types of technologies, are today available on the market. Such an e-nose would have to be the receiver of your odour emissions. Maybe in stereo, with one analyser in each armpit? The analysed message would then have to be turned into digital code and transmitted to the olfactory "loudspeaker" or "TV screen" at the receiving end. And this is where we are really still at a loss. How do you recreate the smell detected at the sender location in a faithful way and deliver it to the receiver?

A few years back I took part in a project funded by the EU program Future and Emerging Technologies (FET). It was again the very agile mind of Noam Sobel from

Chapter 2 that got the project on its feet. The idea behind the programme was to build tuneable odour molecules from DNA snippets. A kind of DNA origami. As mentioned in several of the earlier chapters, the recognition of a smell molecule by the olfactory receptor is thought to happen a bit like a key fitting into a lock. The right molecule somehow fits into the receptor lock. Well, if you could mimic the molecular key by something that looked the same, you should in principle be able to get the receptor lock to recognize it and thereby release the nerve signals telling the brain that a certain odour is present. The idea was thus to create pieces of DNA that would fit into the receptor. The second idea was to fit each DNA molecule with a magnetic part allowing it to be "tuned", so that it would be possible to change the spatial characteristics and thereby the odour impression remotely, maybe with your smartphone.

As you will have noticed by now, this project was very much science fiction. A major obstacle was that DNA molecules are too heavy to float in the air and therefore need to be volatilized somehow. This meant that we could never really test the system. Several other practical problems occurred along the way, as often happens in science, but some interesting insights still came out of the project. Maybe a new wound-healing strategy might emerge from it. But that's a totally different story.

What's the takeaway from this? It's still very hard to create an odour-emitting machine that receives digital information from an analyser and translates it into appropriate olfactory impressions, an olfactometer. For single molecules it can be done, but as discussed earlier, single molecules occur very seldom in nature.

Maybe the real breakthrough will come when we learn how to plug into the olfactory neural network to recreate odour impressions by electrical stimulation, much like we do with cochlear implants in the ear. But this lies far in the future. At the moment, we have no idea how to send odour message via technology. We still have to rely on our other senses via sounds and vision.

What about all the odours we use to manipulate others or to change the smell of ourselves? As mentioned in Chapters 2 and 14, this is a huge and growing market, where millions of dollars are spent on research every year. New mixtures of smells will for sure be invented and sold to us as the latest perfume. Combinations of psychological and olfactory research will also continue to search for the irresistible odours that tempt us to invest in certain merchandise. So much profit relies on these developments that we will for sure see fast changes.

Finally, when it comes to really understanding how we smell, I think that the next quantum leap will occur when we really understand how a molecule is identified by a smell receptor and how all the messages arriving from all the nerves in your nose are intertwined to form the final odour image in your brain.

Acknowledgements

In my career, I've held many popular talks. At these events people in the audience often came up and said that I should write a book telling the stories about smelling and smell. I want to thank all these people for encouraging me to start the project of writing it all up. While writing I have had immense help from Deborah Capras in finding missing information, in formulating parts of the chapters and in making the language flow. Several colleagues, friends and family members have also read the chapters and provided comments that have improved the final results considerably: Susanne Erland, Agnes Erland-Hansson, Otto Erland-Hansson, Manfred Gahr, Jonathan Gershenzon, Rickard Ignell, Markus Knaden, Trese Leinders-Zufall, Johan Lundström, Sigrid Netherer, Silke Sachse, Martin Schroeder and Noam Sobel. I thank them all for reading many pages and providing me with new insights and sometimes correcting me where I was wrong.

Finally, thanks to all colleagues in the field of olfaction, who share my fascination over the chemical analysis that we and other organisms constantly perform of our environment.

Key References

CHAPTER 1

1 Crutzen, P.J. and Stoermer, E.F. (2000) The "Anthropocene". Global Change Newsletter, 41, 17

2 Lindsey, R. (2020) Climate Change: Atmospheric Carbon Dioxide. Climate.gov. Retrieved from https://www.climate.gov/news-features/understanding-climate/climate-change-atmospheric-carbon-dioxide

3 Drake, B. G., Gonzalez-Meler, M. A., & Long, S. P. (1997). MORE EFFICIENT PLANTS: A Consequence of Rising Atmospheric CO2?. Annual review of plant physiology and plant molecular biology, 48, 609–639. https://doi.org/10.1146/annurev.arplant.48.1.609

4 Goyret, Joaquín & Markwell, Poppy & Raguso, Robert. (2008). Context- and scale-dependent effects of floral CO2 on nectar foraging by Manduca sexta. Proceedings of the National Academy of Sciences of the United States of America. 105. 4565-70. 10.1073/pnas.0708629105

5 Majeed, Shahid & Hill, Sharon & Ignell, Rickard. (2013). Impact of elevated CO2 background levels on the host-seeking behaviour of Aedes aegypti. The Journal of experimental biology. 217. 10.1242/jeb.092718

6 Tang, C., Davis, K. E., Delmer, C., Yang, D., & Wills, M. A. (2018). Elevated atmospheric CO2 promoted speciation in mosquitoes (Diptera, Culicidae). Communications biology, 1, 182. https://doi.org/10.1038/s42003- 018-0191-7

7 Haugan PM, Drange H (1996) Effects of CO2 on the ocean environment. Energy Convers Manag 37:1019– 1022 https://doi.org/10.1016/0196-8904(95)00292-8

8 Porteus, Cosima & Hubbard, Peter & Uren Webster, Tamsyn & van Aerle, Ronny & Canario, Adelino & Santos, Eduarda &

Wilson, Rod. (2018). Near-future CO2 levels impair the olfactory system of a marine fish. Nature Climate Change. 8. 10.1038/s41558-018-0224-8

9 Yeung, L. Y., Murray, L. T., Martinerie, P., Witrant, E., Hu, H., Banerjee, A., Orsi, A., & Chappellaz, J. (2019). Isotopic constraint on the twentieth-century increase in tropospheric ozone. Nature, 570(7760), 224– 227. https://doi.org/10.1038/s41586-019-1277-1

10 Seibold, S., Gossner, M.M., Simons, N.K. et al. (2019). Arthropod decline in grasslands and forests is associated with landscape-level drivers. Nature. 574. 671–674. 10.1038/s41586-019-1684-3

11 Cook, B., Haverkamp, A., Hansson, B.S. et al. (2020). Pollination in the Anthropocene: a Moth Can Learn Ozone-Altered Floral Blends. Journal of Chemical Ecology. 1-10. 10.1007/s10886-020-01211-4

12 Girling, R., Lusebrink, I., Farthing, E. et al. (2013). Diesel exhaust rapidly degrades floral odours used by honeybees. Sci Rep 3, 2779. https://doi.org/10.1038/srep02779

13 Kessler, S., Tiedeken, E. J., Simcock, K. L., Derveau, S., Mitchell, J., Softley, S., Stout, J. C., & Wright, G.A. (2015). Bees prefer foods containing neonicotinoid pesticides. Nature, 521(7550), 74–76. https://doi.org/10.1038/nature14414

14 K., Lippi, C. A., Johnson, L. R., Neira, M., Rohr, J. R., Ryan, S. J., Savage, V., Shocket, M. S., Sippy, R., Stewart Ibarra, A. M., Thomas, M. B., & Villena, O. (2019). Thermal biology of mosquito-borne disease. Ecology letters, 22(10), 1690–1708. https://doi.org/10.1111/ele.13335

15 www.ngice.mpg.de

16 Savoca, M., Wohlfeil, M., Ebeler, S., Nevitt, G. (2016). Marine plastic debris emits a keystone infochemical for olfactory foraging seabirds. Science Advances. 2. e1600395-e1600395. 10.1126/sciadv.1600395

17 Our environment is drowning in plastic unenvironment.org Retrieved from: https://www.unenvironment.org/interactive/beat-plastic-pollution/

18 Wilcox, C., Puckridge, M., Schuyler, Q., Townsend, K., Hardesty, B. (2018). A quantitative analysis linking sea turtle mortality and plastic debris ingestion. Scientific Reports. 8. 10.1038/s41598-018-30038-z

19 L. Lebreton, B. Slat, F. Ferrari, B. Sainte-Rose, J. Aitken, R. Marthouse, S. Hajbane, S. Cunsolo, A. Schwarz, A. Levivier, K. Noble, P. Debeljak, H. Maral, R. Schoeneich-Argent, R. Brambini, J. Reisser. (2018). Evidence that the Great Pacific Garbage Patch is rapidly accumulating plastic. Scientific Reports. 2018. 10.1038/ s41598- 018-22939-w

20 Lindeque, Penelope & Cole, Matthew & Coppock, Rachel & Lewis, Ceri & Miller, Rachael & Watts, Andrew & Wilson-McNeal, Alice & Wright, Stephanie & Galloway, Tamara. (2020). Are we underestimating microplastic abundance in the marine environment? A comparison of microplastic capture with nets of different mesh-size. Environmental Pollution. 265. 114721. 10.1016/j. envpol.2020.114721

21 Beyers, D., Farmer, M. (2001). Effects of copper on olfaction of Colorado pikeminnow. Environmental toxicology and chemistry / SETAC. 20. 907-12. 10.1002/etc.5620200427

22 Tierney, K., Sampson, J., Ross, P., Sekela, M., Kennedy, C. (2008). Salmon Olfaction Is Impaired by an Environmentally Realistic Pesticide Mixture. Environmental science & technology. 42. 4996-5001.10.1021/es800240u

23 Ward, A. J., Duff, A. J., Horsfall, J. S., & Currie, S. (2008). Scents and scents-ability: pollution disrupts chemical social recognition and shoaling in fish. Proceedings. Biological sciences, 275(1630), 101–105. https://doi.org/10.1098/rspb.2007.1283

24 Ajmani, G. S., Suh, H. H., & Pinto, J. M. (2016). Effects of Ambient Air Pollution Exposure on Olfaction: A Review. Environmental health perspectives, 124(11), 1683–1693. https://doi.org/10.1289/ EHP136

25 Calderón-Garcidueñas, L., González-Maciel, A., Reynoso-Robles, A., Hammond, J., Kulesza, R., Lachmann, I., Torres-Jardón, R., Mukherjee, P.S., Maher, B.A. (2020) Quadruple abnormal protein aggregates in brainstem pathology and exogenous metal-rich magnetic nanoparticles (and engineered Ti-rich nanorods). The substantia nigrae is a very early target in young urbanites and the gastrointestinal tract a key brainstem portal, Environmental Research, Volume 191, 2020, 110139, ISSN 0013-9351, https://doi. org/10.1016/j.envres.2020.110139

26 Butowt, R., & von Bartheld, C. S. (2020). Anosmia in COVID-19: Underlying Mechanisms and Assessment of an Olfactory Route to Brain Infection. The Neuroscientist : a review journal bringing neurobiology, neurology and psychiatry, 1073858420956905. Advance online publication. https://doi.org/10.1177/1073858420956905

CHAPTER 2

1 https://www.iff.com/

2 Update to Coronavirus symptoms www.gov.scot Retrieved from https://www.gov.scot/news/update-to- coronavirus-symptoms/

3 Stopsack, K. H., Mucci, L. A., Antonarakis, E. S., Nelson, P. S., & Kantoff, P. W. (2020). TMPRSS2 and COVID-19: Serendipity or Opportunity for Intervention?. Cancer discovery, 10(6), 779–782. https://doi.org/10.1158/2159-8290.CD-20-0451

4 Baig AM, Khaleeq A, Ali U, Syeda H. (2020) Evidence of the COVID-19 Virus Targeting the CNS: Tissue Distribution, Host-Virus Interaction, and Proposed Neurotropic Mechanisms. ACS Chemical Neuroscience. 2020 Apr;11(7):995-998. DOI: 10.1021/acschemneuro.0c00122.

5 https://www.mako.co.il/health-news/local/Article-39a265ef1146571026.htm

6 Gilbert, A. (2015) What the Nose Knows: The Science of Scent in Everyday Life, CreateSpace Independent Publishing Platform

7 Bushdid, C., Magnasco, M., Vosshall, L., & Keller, A. (2014). Humans Can Discriminate More than 1 Trillion Olfactory Stimuli. Science, 343(6177), new series, 1370-1372. www.jstor.org/stable/24743486

8 Gerkin, R. C., & Castro, J. B. (2015). The number of olfactory stimuli that humans can discriminate is still unknown. eLife, 4, e08127. https://doi.org/10.7554/eLife.08127

9 Meredith M. (2001). Human vomeronasal organ function: a critical review of best and worst cases. Chemical senses, 26(4), 433–445. https://doi.org/10.1093/chemse/26.4.433

10 Monti-Bloch, L., & Grosser, B. I. (1991). Effect of putative pheromones on the electrical activity of the human vomeronasal organ and olfactory epithelium. The Journal of steroid

biochemistry and molecular biology, 39(4B), 573–582. https://doi.org/10.1016/0960-0760(91)90255-4

11 Savic, I., Berglund, H., Gulyas, B., & Roland, P. (2001). Smelling of odorous sex hormone-like compounds causes sex-differentiated hypothalamic activations in humans. Neuron, 31(4), 661–668. https://doi.org/10.1016/s0896-6273(01)00390-7

12 Savic, I., Berglund, H., & Lindström, P. (2005). Brain response to putative pheromones in homosexual men. Proceedings of the National Academy of Sciences of the United States of America, 102(20), 7356–7361. https://doi.org/10.1073/pnas.0407998102

13 Berglund, Hans & Lindström, Per & Savic, Ivanka. (2006). Brain response to putative pheromones in lesbian women. Proceedings of the National Academy of Sciences of the United States of America. 103. 8269-74.10.1073/pnas.0600331103

14 Wyatt T. D. (2015). The search for human pheromones: the lost decades and the necessity of returning to first principles. Proceedings. Biological sciences, 282(1804), 20142994. https://doi.org/10.1098/rspb.2014.2994

15 Vaglio S. (2009). Chemical communication and mother-infant recognition. Communicative & integrative biology, 2(3), 279–281. https://doi.org/10.4161/cib.2.3.8227

16 Lundström, J. N., Mathe, A., Schaal, B., Frasnelli, J., Nitzsche, K., Gerber, J., & Hummel, T. (2013). Maternal status regulates cortical responses to the body odor of newborns. Frontiers in psychology, 4, 597. https://doi.org/10.3389/fpsyg.2013.00597

17 Uebi, T., Hariyama, T., Suzuki, K., Kanayama, N., Nagata, Y., Ayabe-Kanamura, S., Yanase, S., Ohtsubo, Y., & Ozaki, M. (2019). Sampling, identification and sensory evaluation of odors of a newborn baby's head and amniotic fluid. Scientific reports, 9(1), 12759. https://doi.org/10.1038/s41598-019-49137-6

18 Schaal, B., Marlier, L., & Soussignan, R. (2000). Human foetuses learn odours from their pregnant mother's diet. Chemical senses, 25(6), 729–737. https://doi.org/10.1093/chemse/25.6.729

19 Schicker, I. (2001) For Fathers and Newborns, Natural Law and Odor Retrieved from https://www.washingtonpost.com/archive/politics/2001/02/26/for-fathers-and-newborns-natural-law-and- odor/ccc5982c-acdd-4d0a-8b06-20d2a2bc419a/

20 Chen, D., Katdare, A., & Lucas, N. (2006). Chemosignals of fear enhance cognitive performance in humans. Chemical senses, 31(5), 415–423. https://doi.org/10.1093/chemse/bjj046

21 Gelstein, S., Yeshurun, Y., Rozenkrantz, L., Shushan, S., Frumin, I., Roth, Y., & Sobel, N. (2011). Human tears contain a chemosignal. Science (New York, N.Y.), 331(6014), 226–230. https://doi.org/10.1126/science.1198331

22 Oh, T. J., Kim, M. Y., Park, K. S., & Cho, Y. M. (2012). Effects of chemosignals from sad tears and postprandial plasma on appetite and food intake in humans. PloS one, 7(8), e42352. https://doi.org/10.1371/journal.pone.0042352

23 Ferrero, D. M., Moeller, L. M., Osakada, T., Horio, N., Li, Q., Roy, D. S., Cichy, A., Spehr, M., Touhara, K., & Liberles, S. D. (2013). A juvenile mouse pheromone inhibits sexual behaviour through the vomeronasal system. Nature, 502(7471), 368–371. https://doi.org/10.1038/nature12579

24 Keller, A., Zhuang, H., Chi, Q., Vosshall, L., Matsunami, H. (2007). Genetic Variation in a Human Odorant Receptor Alters Odour Perception. Nature. 449. 468-72. 10.1038/nature06162

25 Wedekind, C., Seebeck, T., Bettens, F., & Paepke, A. J. (1995). MHC-dependent mate preferences in humans. Proceedings. Biological sciences, 260(1359), 245–249. https://doi.org/10.1098/rspb.1995.0087

26 Milinski, M., Croy, I., Hummel, T., & Boehm, T. (2013). Major histocompatibility complex peptide ligands as olfactory cues in human body odour assessment. Proceedings of the Royal Society B: Biological Sciences, 280(1757), 20130381. https://doi.org/10.1098/rspb.2013.0381

27 McClintock, M. (1971) M. Menstrual Synchrony and Suppression. Nature 229, 244–245 (1971). https://doi.org/10.1038/229244a0

28 Russell, M. J., Switz, G. M., & Thompson, K. (1980). Olfactory influences on the human menstrual cycle. Pharmacology, biochemistry, and behavior, 13(5), 737–738. https://doi.org/10.1016/0091-3057(80)90020-9

29 Stern, K., & McClintock, M. K. (1998). Regulation of ovulation by human pheromones. Nature, 392(6672), 177–179. https://doi.org/10.1038/32408

30 Ziomkiewicz A. (2006). Menstrual synchrony: Fact or artifact?. Human nature (Hawthorne, N.Y.), 17(4), 419–432. https://doi.org/10.1007/s12110-006-1004-0

31 Åhs, F., Miller, S., Gordon, A., & Lundström, J. (2013). Aversive learning increases sensory detection sensitivity. Biological Psychology, 92, 135-141

32 Sinding, C., Valadier, F., Al-Hassani, V., Feron, G., Tromelin, A., Kontaris, I., & Hummel, T. (2017). New determinants of olfactory habituation. Scientific Reports, 7

33 Khan, R. M., Luk, C. H., Flinker, A., Aggarwal, A., Lapid, H., Haddad, R., & Sobel, N. (2007). Predicting odor pleasantness from odorant structure: pleasantness as a reflection of the physical world. The Journal of neuroscience : the official journal of the Society for Neuroscience, 27(37), 10015–10023. https://doi.org/10.1523/JNEUROSCI.1158-07.2007

34 Ravia, A., Snitz, K., Honigstein, D., Finkel, M., Zirler, R., Perl, O., Secundo, L., Laudamiel, C., Harel, D., & Sobel, N. (2020). A measure of smell enables the creation of olfactory metamers. Nature, 10.1038/s41586-020- 2891-7. Advance online publication. https://doi.org/10.1038/s41586-020-2891-7

35 Olofsson, J. K., Hurley, R. S., Bowman, N. E., Bao, X., Mesulam, M. M., & Gottfried, J. A. (2014). A designated odor-language integration system in the human brain. The Journal of neuroscience: the official journal of the Society for Neuroscience, 34(45), 14864–14873. https://doi.org/10.1523/JNEUROSCI.2247- 14.2014

36 Majid, A., Burenhult, N., Stensmyr, M., de Valk, J., & Hansson, B. S. (2018). Olfactory language and abstraction across cultures. Philosophical transactions of the Royal Society of London. Series B, Biological sciences, 373(1752), 20170139. https://doi.org/10.1098/rstb.2017.0139

CHAPTER 3

1 Walker, D., Walker, J., Cavnar, P., Taylor, J., Pickel, D., Hall, S., & Suarez, J. (2006). Naturalistic quantification of canine olfactory sensitivity. Applied animal behaviour science, 97, 241-254. doi: 10.1016/j.applanim.2005.07.009

2 Kester, D., Settles, G. (1998). The External Aerodynamics of Canine Olfaction. -1. 10.1007/978-3-7091-6025- 1_23
3 Jenkins, E. K., DeChant, M. T., & Perry, E. B. (2018). When the Nose Doesn't Know: Canine Olfactory Function Associated With Health, Management, and Potential Links to Microbiota. Frontiers in veterinary science, 5, 56. https://doi.org/10.3389/fvets.2018.00056
4 Glausiusz, J. (2008). The Hidden Power of SCENT. Scientific American Mind, 19(4), 38-45. Retrieved November 14, 2020, from http://www.jstor.org/stable/24939934
5 Horowitz, A. (2015). Reading Dogs Reading Us. Proceedings of the American Philosophical Society, 159(2), 141-155. Retrieved November 14, 2020, from http://www.jstor.org/stable/24640211
6 Botigué, L., Song, S., Scheu, A. et al. (2017) Ancient European dog genomes reveal continuity since the Early Neolithic. Nat Commun 8, 16082 (2017) doi:10.1038/ncomms16082
7 Gadbois, S., & Reeve, C. (2014) Chapter 1 Canine Olfaction: Scent, Sign, and Situation
8 Nagasawa, M., Mitsui, S., En, S., Ohtani, N., Ohta, M., Sakuma, Y., Onaka, T., Mogi, K., & Kikusui, T. (2015). Oxytocin-gaze positive loop and the coevolution of human-dog bonds. Science, 348, 333 - 336
9 Wells, D., & Hepper, P. (2003). Directional tracking in the domestic dog, Canis familiaris. Applied Animal Behaviour Science, 84, 297-305
10 Hepper, P., Wells, D. (2005). How many footsteps do dogs need to determine the direction of an odour trail? Chemical Senses, 30(4) (4), 291-298. https://doi.org/10.1093/chemse/bji023
11 Akpan, N., Ehrichs, M. (2016) Inside the extraordinary nose of a search-and-rescue dog PBS News Hour. Retrieved from https://www.pbs.org/newshour/science/inside-nose-rescue-dog

CHAPTER 4

1 Krulwich, R., (2014) What Not To Serve Buzzards For Lunch, A Glorious Science Experiment. NPR.org Retrieved from https://www.npr.org/sections/krulwich/2014/06/26/325648459/what-not-to-serve-buzzards-for- lunch-a-glorious-science-experiment

2 Houston, D.C. (1986) Scavenging Efficiency of Turkey Vultures in Tropical Forest, The Condor, Volume 88, Issue 3, 1 August 1986, Pages 318–323, https://doi.org/10.2307/1368878

3 Grigg, N.P., Krilow, J.M., Gutiérrez-Ibáñez, C., Wylie, D.R., Graves, G., & Iwaniuk, A. (2017). Anatomical evidence for scent guided foraging in the turkey vulture. Scientific Reports, 7

4 Averett, N. (2014) Birds Can Smell, and One Scientist is Leading the Charge to Prove It Audubon.org Retrieved at https://www.audubon.org/magazine/january-february-2014/birds-can-smell-and-one-scientist

5 Bonadonna, F., Bajzak, C., Benhamou, S., Igloi, K., Jouventin, P., Lipp, H. P., & Dell'Omo, G. (2005). Orientation in the wandering albatross: interfering with magnetic perception does not affect orientation performance. Proceedings. Biological sciences, 272(1562), 489–495. https://doi.org/10.1098/rspb.2004.2984

6 Gagliardo, A., Bried, J., Lambardi, P., Luschi, P., Wikelski, M., & Bonadonna, F. (2013). Oceanic navigation in Cory's shearwaters: evidence for a crucial role of olfactory cues for homing after displacement. The Journal of experimental biology, 216(Pt 15), 2798–2805. https://doi.org/10.1242/jeb.085738

7 Reynolds, A., Cecere, J.G., Paiva, V., Ramos, J., & Focardi, S. (2015). Pelagic seabird flight patterns are consistent with a reliance on olfactory maps for oceanic navigation. Proceedings of the Royal Society B: Biological Sciences, 282

8 Mardon, J., Nesterova, A. P., Traugott, J., Saunders, S. M., & Bonadonna, F. (2010). Insight of scent: experimental evidence of olfactory capabilities in the wandering albatross (Diomedea exulans). The Journal of experimental biology, 213(4), 558–563. https://doi.org/10.1242/jeb.032979

9 Peps, S. (1666) The Diary of Samuel Pepys, Sunday 2 September 1666, Retrieved from https://www.pepysdiary.com/diary/1666/09/02/

10 Reuters, (2008) Chronology: Reuters, from pigeons to multimedia merger Retrieved from https://www.reuters.com/article/us-reuters-thomson-chronology/chronology-reuters-from-pigeons-to- multimedia-merger-idUSL1849100620080219

11 Corera, G. (2018) Operation Columba: The Secret Pigeon Service:

The Untold Story of World War II Resistance in Europe, William Morrow; Illustrated edition (October 16, 2018)

12 Wallraff H.G, (2005) Avian Navigation: Pigeon Homing as a Paradigm Springer, Berlin

13 Caro, S. P., & Balthazart, J. (2010). Pheromones in birds: myth or reality?. Journal of comparative physiology. A, Neuroethology, sensory, neural, and behavioral physiology, 196(10), 751–766. https://doi.org/10.1007/s00359-010-0534-4

14 Gagliardo, A., Pollonara, E., & Wikelski, M. (2016). Pigeon navigation: exposure to environmental odours prior to release is sufficient for homeward orientation, but not for homing. The Journal of experimental biology, 219(Pt 16), 2475–2480. https://doi.org/10.1242/jeb.140889

15 Lengagne, T., Jouventin, P., & Aubin, T. (1999). Finding One's Mate in a King Penguin Colony: Efficiency of Acoustic Communication. Behaviour, 136(7), 833-846. Retrieved November 14, 2020, from http://www.jstor.org/stable/4535644

16 Birds' Sense of Smell. (2011). The Science Teacher, 78(8), 24-27. Retrieved November 14, 2020, from http://www.jstor.org/stable/24148500

17 Krause, E. T., Krüger, O., Kohlmeier, P., & Caspers, B. A. (2012). Olfactory kin recognition in a songbird. Biology letters, 8(3), 327–329. https://doi.org/10.1098/rsbl.2011.1093

18 Caspers, B. A., Hagelin, J. C., Paul, M., Bock, S., Willeke, S., & Krause, E. T. (2017). Zebra Finch chicks recognise parental scent, and retain chemosensory knowledge of their genetic mother, even after egg cross- fostering. Scientific reports, 7(1), 12859. https://doi.org/10.1038/s41598-017-13110-y

19 Whittaker, D. J., Slowinski, S. P., Greenberg, J. M., Alian, O., Winters, A. D., Ahmad, M. M., Burrell, M., Soini, H. A., Novotny, M. V., Ketterson, E. D., & Theis, K. R. (2019). Experimental evidence that symbiotic bacteria produce chemical cues in a songbird. The Journal of experimental biology, 222(Pt 20), jeb202978. https://doi.org/10.1242/jeb.202978

20 Caro, S. P., & Balthazart, J. (2010). Pheromones in birds: myth or reality?. Journal of comparative physiology. A, Neuroethology, sensory, neural, and behavioral physiology, 196(10), 751–766. https://doi.org/10.1007/s00359-010-0534-4

21 Steiger, S. S., Fidler, A. E., Valcu, M., & Kempenaers, B. (2008). Avian olfactory receptor gene repertoires: evidence for a well-developed sense of smell in birds?. Proceedings. Biological sciences, 275(1649), 2309– 2317. https://doi.org/10.1098/rspb.2008.0607
22 Meteyer, C. U., Rideout, B. A., Gilbert, M., Shivaprasad, H. L., & Oaks, J. L. (2005). Pathology and proposed pathophysiology of diclofenac poisoning in free-living and experimentally exposed oriental white-backed vultures (Gyps bengalensis). Journal of wildlife diseases, 41(4), 707–716. https://doi.org/10.7589/0090-3558- 41.4.707
23 Savoca, M. S., Wohlfeil, M. E., Ebeler, S. E., & Nevitt, G. A. (2016). Marine plastic debris emits a keystone infochemical for olfactory foraging seabirds. Science advances, 2(11), e1600395. https://doi.org/10.1126/sciadv.1600395

CHAPTER 5

1 Catania K. C. (2006). Olfaction: underwater 'sniffing' by semi-aquatic mammals. Nature, 444(7122), 1024– 1025. https://doi.org/10.1038/4441024a
2 Reiten, I., Uslu, F.E., Fore, S., Pelgrims, R., Ringers, C., Verdugoa, C.D., Hoffmann, M., Lal, P., Kawakami, K., Pekkan, K., et al. (2017). Motile-cilia-mediated flow improves sensitivity and temporal resolution of olfactory computations. Curr. Biol. 27, 166–174. https://www.cell.com/current-biology/fulltext/S0960- 9822(16)31389-6
3 Neuhauss S. C. (2017). Olfaction: How Fish Catch a Whiff. Current biology: CB, 27(2), R57–R58. https://doi.org/10.1016/j.cub.2016.12.007
4 Hamdani, e., Døving, K. B. (2007). The functional organization of the fish olfactory system. Progress in neurobiology, 82(2), 80–86. https://doi.org/10.1016/j.pneurobio.2007.02.007
5 Stacey, N., Sorensen, P. (2002). Hormonal Pheromones in Fish. 10.1016/B978-008088783-8.00018-8
6 Jumper, G., Baird, R. (1991). Location by Olfaction: A Model and Application to the Mating Problem in the Deep-Sea Hatchetfish Argyropelecus hemigymnus. The American Naturalist, 138(6), 1431-1458. Retrieved October 27, 2020, from http://www.jstor.org/stable/2462555

7 Vieira, S., Biscoito, M., Encarnação, H., Delgado, J., & Pietsch, T. (2013). Sexual Parasitism in the Deep-sea Ceratioid Anglerfish Centrophryne spinulosa Regan and Trewavas (Lophiiformes: Centrophrynidae). Copeia, 2013(4), 666-669. Retrieved September 21, 2020, from http://www.jstor.org/stable/24637159

8 Pietsch, T. (2009). Oceanic Anglerfishes: Extraordinary Diversity in the Deep Sea. University of California Press. Retrieved November 14, 2020, from http://www.jstor.org/stable/10.1525/j.ctt1ppb32 pp. 43-45 e

9 NOAA. (2019) What is a sea lamprey? Retrieved from https://oceanservice.noaa.gov/facts/sea-lamprey.html

10 Johnson, N., Yun, S., Thompson, H., Brant, C., Li, W., & Meinwald, J. (2009). A Synthesized Pheromone Induces Upstream Movement in Female Sea Lamprey and Summons Them into Traps. Proceedings of the National Academy of Sciences of the United States of America, 106(4), 1021-1026. www.jstor.org/stable/40254676

11 Li, W., Scott, A. P., Siefkes, M. J., Yan, H., Liu, Q., Yun, S. S., & Gage, D. A. (2002). Bile Acid secreted by male sea lamprey that acts as a sex pheromone. Science (New York, N.Y.), 296(5565), 138–141. https://doi.org/10.1126/science.1067797

12 Bandoh, H., Kida, I., & Ueda, H. (2011). Olfactory responses to natal stream water in sockeye salmon by BOLD fMRI. PloS one, 6(1), e16051. https://doi.org/10.1371/journal.pone.0016051

13 Roberts, L., Garcia de Leaniz, C. (2011). Something smells fishy: Predator-naïve salmon use diet cues, not kairomones, to recognize a sympatric mammalian predator. Animal Behaviour. 82. 619-625.10.1016/j.anbehav.2011.06.019

14 Brooker, R. M., Munday, P. L., Chivers, D. P., & Jones, G. P. (2015). You are what you eat: diet-induced chemical crypsis in a coral-feeding reef fish. Proceedings. Biological sciences, 282(1799), 20141887. https://doi.org/10.1098/rspb.2014.1887

15 Gardiner, J. M., Whitney, N. M., & Hueter, R. E. (2015). Smells Like Home: The Role of Olfactory Cues in the Homing Behavior of Blacktip Sharks, Carcharhinus limbatus. Integrative and comparative biology, 55(3), 495–506. https://doi.org/10.1093/icb/icv087

16 Marks, R. In-depth: Shark Senses PBS.org/ Retrieved from https://www.pbs.org/kqed/oceanadventures/episodes/sharks/indepth-senses.html

17 Gardiner, J. M., & Atema, J. (2010). The function of bilateral odor arrival time differences in olfactory orientation of sharks. Current biology: CB, 20(13), 1187–1191. https://doi.org/10.1016/j.cub.2010.04.053

18 Enjin, A., & Suh, G. S. (2013). Neural mechanisms of alarm pheromone signaling. Molecules and cells, 35(3), 177–181. https://doi.org/10.1007/s10059-013-0056-3

19 Mathuru, A. S., Kibat, C., Cheong, W. F., Shui, G., Wenk, M. R., Friedrich, R. W., & Jesuthasan, S. (2012). Chondroitin fragments are odorants that trigger fear behavior in fish. Current biology: CB, 22(6), 538–544. https://doi.org/10.1016/j.cub.2012.01.061

20 Walker, M. (2010) Whale 'sense of smell' revealed, BBC Earth News Retrieved from http://news.bbc.co.uk/earth/hi/earth_news/newsid_8844000/8844443.stm

21 George, J.C., Thewissen, H. Bowhead Whale Sensory Research / Olfaction in Bowhead Whales North- Slope.org Retrieved from http://www.north-slope.org/departments/wildlife-management/studies-and-research- projects/bowhead-whales/bowhead-whale-anatomy-and-physiology-studies/bowhead-whale-sensory- research#OlfactionBH

22 Pitcher, B.J., Harcourt, R., Schaal, B., & Charrier, I. (2010). Social olfaction in marine mammals: wild female Australian sea lions can identify their pup's scent. Biology Letters, 7, 60 - 62

23 Stoffel, M., Caspers, B.A., Forcada, J., Giannakara, A., Baier, M., Eberhart-Phillips, L., Müller, C., & Hoffman, J.I. (2015). Chemical fingerprints encode mother–offspring similarity, colony membership, relatedness, and genetic quality in fur seals. Proceedings of the National Academy of Sciences, 112, E5005 - E5012

Chapter 6

1 Schröder, H., Moser, N., Huggenberger, S. (2020) Neuroanatomy of the Mouse: An introduction, pp 319-331 The Mouse Olfactory System, Springer International Publishing https://www.springer.com/gp/book/9783030198978

2 Zhang, X., Firestein, S. (2002). The olfactory receptor gene superfamily of the mouse. Nature neuroscience, 5(2), 124–133. https://doi.org/10.1038/nn800

3 Mombaerts, P. (1996) Targeting olfaction, Current Opinion in Neurobiology, Volume 6, Issue 4,1996, Pages 481-486, ISSN

0959-4388, https://doi.org/10.1016/S0959-4388(96)80053-5. (http://www.sciencedirect.com/science/article/pii/S0959438896800535)

4 Mombaerts, P. (2006). Axonal wiring in the mouse olfactory system. Annual review of cell and developmental biology, 22, 713-37

5 Zancanaro, C. (2014) Vomeronasal Organ: A Short History of Discovery and an Account of Development and Morphology in the Mouse. In: Mucignat-Caretta C, editor. Neurobiology of Chemical Communication. Boca Raton (FL): CRC Press/Taylor & Francis; 2014. Chapter 9. Retrieved from: https://www.ncbi.nlm.nih.gov/books/NBK200982/

6 Pérez-Gómez, A., Stein, B., Leinders-Zufall, T., & Chamero, P. (2014). Signaling mechanisms and behavioral function of the mouse basal vomeronasal neuroepithelium. Frontiers in neuroanatomy, 8, 135. https://doi.org/10.3389/fnana.2014.00135

7 Fleischer, J., & Breer, H. (2010). The Grueneberg ganglion: a novel sensory system in the nose. Histology and histopathology, 25(7), 909–915. https://doi.org/10.14670/HH-25.909

8 Brechbühl, J., Vallière, A., Wood, D., Nenniger Tosato, M., Broillet, M. (2020). The Grueneberg ganglion controls odor-driven food choices in mice under threat. Communications Biology. 3. 10.1038/s42003-020- 01257-w

9 Brechbühl, J., Klaey, M., Broillet, M. C. (2008). Grueneberg ganglion cells mediate alarm pheromone detection in mice. Science (New York, N.Y.), 321(5892), 1092–1095. https://doi.org/10.1126/science.1160770

10 Schmid, A., Pyrski, M., Biel, M., Leinders-Zufall, T., & Zufall, F. (2010). Grueneberg ganglion neurons are finely tuned cold sensors. The Journal of neuroscience : the official journal of the Society for Neuroscience, 30(22), 7563–7568. https://doi.org/10.1523/JNEUROSCI.0608-10.2010

11 Barrios, A. W., Núñez, G., Sánchez Quinteiro, P., Salazar, I. (2014). Anatomy, histochemistry, and immunohistochemistry of the olfactory subsystems in mice. Frontiers in neuroanatomy, 8, 63. https://doi.org/10.3389/fnana.2014.00063

12 Ma, M., Grosmaitre, X., Iwema, C. L., Baker, H., Greer, C. A., Shepherd, G. M. (2003). Olfactory signal transduction in the mouse

septal organ. The Journal of neuroscience : the official journal of the Society for Neuroscience, 23(1), 317–324. https://doi.org/10.1523/JNEUROSCI.23-01-00317.2003

13 Tian, H., Ma, M. (2004). Molecular Organization of the Olfactory Septal Organ. The Journal of neuroscience: the official journal of the Society for Neuroscience. 24. 8383-90. 10.1523/JNEUROSCI.2222-04.2004

14 Liberles S. D. (2014). Mammalian pheromones. Annual review of physiology, 76, 151–175. https://doi.org/10.1146/annurev-physiol-021113-170334

15 Chamero, P., Marton, T. F., Logan, D. W., Flanagan, K., Cruz, J. R., Saghatelian, A., Cravatt, B. F., & Stowers, L. (2007). Identification of protein pheromones that promote aggressive behaviour. Nature, 450(7171), 899–902. https://doi.org/10.1038/nature05997

16 Novotny, M., Harvey, S., Jemiolo, B., & Alberts, J. (1985). Synthetic pheromones that promote inter-male aggression in mice. Proceedings of the National Academy of Sciences of the United States of America, 82(7), 2059–2061. https://doi.org/10.1073/pnas.82.7.2059

17 Logan, D. W., Brunet, L. J., Webb, W. R., Cutforth, T., Ngai, J., & Stowers, L. (2012). Learned recognition of maternal signature odors mediates the first suckling episode in mice. Current biology : CB, 22(21), 1998– 2007. https://doi.org/10.1016/j.cub.2012.08.041

18 Roberts, S. A., Simpson, D. M., Armstrong, S. D., Davidson, A. J., Robertson, D. H., McLean, L., Beynon, R. J., & Hurst, J. L. (2010). Darcin: a male pheromone that stimulates female memory and sexual attraction to an individual male's odour. BMC biology, 8, 75. https://doi.org/10.1186/1741-7007-8-75

19 Bruce, H. M. (1959). An exteroceptive block to pregnancy in the mouse. Nature, 184, 105. https://doi.org/10.1038/184105a0

20 Whitten, W. K. (1959). Occurrence of anoestrus in mice caged in groups. The Journal of endocrinology, 18(1), 102–107. https://doi.org/10.1677/joe.0.0180102

21 Vandenbergh J. G. (1969). Male odor accelerates female sexual maturation in mice. Endocrinology, 84(3), 658–660. https://doi.org/10.1210/endo-84-3-658

22 Ferrero, D., Lemon, J., Fluegge, D., Pashkovski, S., Korzan, W., Datta, S., Spehr, M., Fendt, M., Liberles, S. (2011). Detection and

avoidance of a carnivore odor by prey. Proceedings of the National Academy of Sciences of the United States of America. 108. 11235-40. 10.1073/pnas.1103317108

23 Dewan, A., Pacifico, R., Zhan, R., Rinberg, D., Bozza, T. (2013). Non-redundant coding of aversive odours in the main olfactory pathway. Nature, 497(7450), 486–489. https://doi.org/10.1038/nature12114

CHAPTER 7

1 Angioy, A. M., Desogus, A., Barbarossa, I. T., Anderson, P., & Hansson, B. S. (2003). Extreme sensitivity in an olfactory system. Chemical senses, 28(4), 279–284. https://doi.org/10.1093/chemse/28.4.279

2 Kaissling, K. E. (2009) The Sensitivity of the Insect Nose: The Example of Bombyx Mori. In: Gutiérrez A., Marco S. (eds) Biologically Inspired Signal Processing for Chemical Sensing. Studies in Computational Intelligence, vol 188. Springer, Berlin, Heidelberg. https://doi.org/10.1007/978-3-642-00176-5_3

3 Karlson, P., Luscher, M. (1959). "Pheromones": a new term for a class of biologically active substances. Nature, 183(4653), 55–56. https://doi.org/10.1038/183055a0

4 Hansson, B.S. Olfaction in Lepidoptera. Experientia 51, 1003–1027 (1995). https://doi.org/10.1007/BF01946910

5 Missbach, C., Dweck, H. K., Vogel, H., Vilcinskas, A., Stensmyr, M. C., Hansson, B. S., & Grosse-Wilde, E. (2014). Evolution of insect olfactory receptors. eLife, 3, e02115. https://doi.org/10.7554/eLife.02115

6 Fatouros, N., Huigens, M., van Loon, J. et al. Butterfly anti-aphrodisiac lures parasitic wasps. Nature 433, 704 (2005). https://doi.org/10.1038/433704a

7 Jones, A. G., & Ratterman, N. L. (2009). Mate choice and sexual selection: what have we learned since Darwin?. Proceedings of the National Academy of Sciences of the United States of America, 106 Suppl 1(Suppl 1), 10001–10008. https://doi.org/10.1073/pnas.0901129106

8 Fisher R. A. (1915). The evolution of sexual preference. The Eugenics review, 7(3), 184–192

9 Edwards A. W. (2000). The genetical theory of natural selection. Genetics, 154(4), 1419–1426

10 ter Hofstede HM, Goerlitz HR, Ratcliffe JM, Holderied MW, Surlykke A. (2013) The simple ears of noctuoid moths are tuned to the calls of their sympatric bat community. J Exp Biol. 2013 Nov 1;216(Pt 21):3954-62. doi: 10.1242/jeb.093294. Epub 2013 Aug 2. PMID: 23913945

11 Svensson GP, Löfstedt C, Skals N. (2007) Listening in pheromone plumes: Disruption of olfactory-guided mateattraction in a moth by a bat-like ultrasound. 9pp.Journal of Insect Science7:59, available online:insectscience.org/7.59

12 Gemeno, C., Yeargan, K.V. & Haynes, K.F. (2000) Aggressive Chemical Mimicry by the Bolas Spider Mastophora hutchinsoni: Identification and Quantification of a Major Prey's Sex Pheromone Components in the Spider's Volatile Emissions. J Chem Ecol 26, 1235–1243 (2000). https://doi.org/10.1023/A:1005488128468

13 Peter Karlson and Adolf Butenandt (1959) Pheromones (Ectohormones) in Insects Annual Review of Entomology 1959 4:1, 39-58 https://doi.org/10.1146/annurev.en.04.010159.000351

14 A. Butenandt, E. Hecker: (1961) Synthese des Bombykols, des Sexuallockstoffes des Seidenspinners, und seiner geometrischen Isomeren. In: Angew. Chem. 73, 1961, S. 349. https://doi.org/10.1002/ange.19610731102

15 Allison, J., Cardé, R. (Eds.). (2016). Pheromone Communication in Moths: Evolution, Behavior, and Application. Oakland, California: University of California Press. Retrieved November 15, 2020, from http://www.jstor.org/stable/10.1525/j.ctv1xxxzm

16 Baker T.C., Vickers N.J. (1997) Pheromone-Mediated Flight in Moths. In: Cardé R.T., Minks A.K. (eds) Insect Pheromone Research. Springer, Boston, MA. https://doi.org/10.1007/978-1-4615-6371-6_23

17 Phelan, PL. (1992) Evolution of sex pheromones and the role of asymmetric tracking. In: Insect chemical ecology: an evolutionary approach, edited Roitberg, B., Isman, M. 1992 Chapman and Hall ISBN: 9780412018718

18 Hansson, B. S., Tóth, M., Löfstedt, C., Szöcs, G., Subchev, M., & Löfqvist, J. (1990). Pheromone variation among eastern European

and a western Asian population of the turnip moth Agrotis segetum. Journal of chemical ecology, 16(5), 1611–1622. https://doi.org/10.1007/BF01014094

19 Wunderer, H., Hansen, K., Bell, T. W., Schneider, D., & Meinwald, J. (1986). Sex pheromones of two Asian moths (Creatonotos transiens, C. gangis; Lepidoptera--Arctiidae): behavior, morphology, chemistry and electrophysiology. Experimental biology, 46(1), 11–27

20 Boppré, M., Schneider, D. (1985) Pyrrolizidine alkaloids quantitatively regulate both scent organ morphogenesis and pheromone biosynthesis in male Creatonotos moths (Lepidoptera: Arctiidae). J. Comp. Physiol. 157, 569–577 (1985). https://doi.org/10.1007/BF01351351

21 Kessler, D., Gase, K., & Baldwin, I. T. (2008). Field experiments with transformed plants reveal the sense of floral scents. Science (New York, N.Y.), 321(5893), 1200–1202. https://doi.org/10.1126/science.1160072

22 Haverkamp, A., Yon, F., Keesey, I. W., Mißbach, C., Koenig, C., Hansson, B. S., Baldwin, I. T., Knaden, M., & Kessler, D. (2016). Hawkmoths evaluate scenting flowers with the tip of their proboscis. eLife, 5, e15039. https://doi.org/10.7554/eLife.15039

CHAPTER 8

1 Hansson, B. S., Knaden, M., Sachse, S., Stensmyr, M. C., & Wicher, D. (2010). Towards plant-odor-related olfactory neuroethology in Drosophila. Chemoecology, 20(2), 51–61. https://doi.org/10.1007/s00049-009-0033- 7

2 Morgan T. H. (1910). SEX LIMITED INHERITANCE IN DROSOPHILA. Science (New York, N.Y.), 32(812), 120–122. https://doi.org/10.1126/science.32.812.120

3 Bellen, H., Tong, C. & Tsuda, H. (2010) 100 years of Drosophila research and its impact on vertebrate neuroscience: a history lesson for the future. Nat Rev Neurosci 11, 514–522 (2010). https://doi.org/10.1038/nrn2839

4 Hansson, B. S., & Stensmyr, M. C. (2011). Evolution of insect olfaction. Neuron, 72(5), 698–711. https://doi.org/10.1016/j.neuron.2011.11.003

5 Stocker R. F. (2009). The olfactory pathway of adult and larval Drosophila: conservation or adaptation to stage-specific needs?. Annals of the New York Academy of Sciences, 1170, 482–486. https://doi.org/10.1111/j.1749-6632.2009.03896.x

6 Vosshall, L. B., Amrein, H., Morozov, P. S., Rzhetsky, A., & Axel, R. (1999). A spatial map of olfactory receptor expression in the Drosophila antenna. Cell, 96(5), 725–736. https://doi.org/10.1016/s0092- 8674(00)80582-6

7 Vosshall, L. B., Wong, A. M., & Axel, R. (2000). An olfactory sensory map in the fly brain. Cell, 102(2), 147– 159. https://doi.org/10.1016/s0092-8674(00)00021-0

8 Dweck, H. K., Ebrahim, S. A., Khallaf, M. A., Koenig, C., Farhan, A., Stieber, R., Weißflog, J., Svatoš, A., Grosse-Wilde, E., Knaden, M., & Hansson, B. S. (2016). Olfactory channels associated with the Drosophila maxillary palp mediate short- and long-range attraction. eLife, 5, e14925. https://doi.org/10.7554/eLife.14925

9 Wicher, Dieter & Schäfer, Ronny & Bauernfeind, René & Stensmyr, Marcus & Heller, Regine & Heinemann, Stefan & Hansson, Bill. (2008). Drosophila odorant receptors are both ligand-gated and cyclic-nucleotide- activated cation channels. Nature. 452. 1007-11. 10.1038/nature06861

10 Sato, K., Pellegrino, M., Nakagawa, T., Nakagawa, T., Vosshall, L. B., & Touhara, K. (2008). Insect olfactory receptors are heteromeric ligand-gated ion channels. Nature, 452(7190), 1002–1006. https://doi.org/10.1038/nature06850

11 Getahun, M. N., Olsson, S. B., Lavista-Llanos, S., Hansson, B. S., & Wicher, D. (2013). Insect odorant response sensitivity is tuned by metabotropically autoregulated olfactory receptors. PloS one, 8(3), e58889. https://doi.org/10.1371/journal.pone.0058889

12 Stensmyr, M. C., Dweck, H. K., Farhan, A., Ibba, I., Strutz, A., Mukunda, L., Linz, J., Grabe, V., Steck, K., Lavista-Llanos, S., Wicher, D., Sachse, S., Knaden, M., Becher, P. G., Seki, Y., & Hansson, B. S. (2012). A conserved dedicated olfactory circuit for detecting harmful microbes in Drosophila. Cell, 151(6), 1345–1357. https://doi.org/10.1016/j.cell.2012.09.046

13 Ebrahim, S. A., Dweck, H. K., Stökl, J., Hofferberth, J. E., Trona, F., Weniger, K., Rybak, J., Seki, Y., Stensmyr, M. C., Sachse,

S., Hansson, B. S., & Knaden, M. (2015). Drosophila Avoids Parasitoids by Sensing Their Semiochemicals via a Dedicated Olfactory Circuit. PLoS biology, 13(12), e1002318. https://doi.org/10.1371/journal.pbio.1002318

14 Dweck, H. K., Ebrahim, S. A., Kromann, S., Bown, D., Hillbur, Y., Sachse, S., Hansson, B. S., & Stensmyr, M. C. (2013). Olfactory preference for egg laying on citrus substrates in Drosophila. Current biology : CB, 23(24), 2472–2480. https://doi.org/10.1016/j.cub.2013.10.047

15 Ejima A. (2015). Pleiotropic actions of the male pheromone cis-vaccenyl acetate in Drosophila melanogaster. Journal of comparative physiology. A, Neuroethology, sensory, neural, and behavioral physiology, 201(9), 927– 932. https://doi.org/10.1007/s00359-015-1020-9

16 Dekker, T., Ibba, I., Siju, K. P., Stensmyr, M. C., & Hansson, B. S. (2006). Olfactory shifts parallel superspecialism for toxic fruit in Drosophila melanogaster sibling, D. sechellia. Current biology : CB, 16(1), 101–109. https://doi.org/10.1016/j.cub.2005.11.075

17 Auer, T. O., Khallaf, M. A., Silbering, A. F., Zappia, G., Ellis, K., Álvarez-Ocaña, R., Arguello, J. R., Hansson, B. S., Jefferis, G., Caron, S., Knaden, M., & Benton, R. (2020). Olfactory receptor and circuit evolution promote host specialization. Nature, 579(7799), 402–408. https://doi.org/10.1038/s41586-020-2073-7

18 Lavista-Llanos, S., Svatoš, A., Kai, M., Riemensperger, T., Birman, S., Stensmyr, M. C., & Hansson, B. S. (2014). Dopamine drives Drosophila sechellia adaptation to its toxic host. eLife, 3, e03785. https://doi.org/10.7554/eLife.03785

19 Keesey, I. W., Knaden, M., & Hansson, B. S. (2015). Olfactory specialization in Drosophila suzukii supports an ecological shift in host preference from rotten to fresh fruit. Journal of chemical ecology, 41(2), 121–128. https://doi.org/10.1007/s10886-015-0544-3

20 Cloonan, K. R., Abraham, J., Angeli, S., Syed, Z., & Rodriguez-Saona, C. (2018). Advances in the Chemical Ecology of the Spotted Wing Drosophila (Drosophila suzukii) and its Applications. Journal of chemical ecology, 44(10), 922–939. https://doi.org/10.1007/s10886-018-1000-y

21 Green, J.E., Cavey, M., Caturegli, E., Gompel, N., Prud'homme, B. (2019) Evolution of ovipositor length in Drosophila suzukii is

driven by enhanced cell size expansion and anisotropic tissue reorganization Current Biology 29:2075–2082. https://doi.org/10.1016/j.cub.2019.05.020

CHAPTER 9

1 Malaria (2020) World Health Organization (Published 2020 Jan 14, Accessed 2020 Nov 16) Retrieved from https://www.who.int/news-room/fact-sheets/detail/malaria
2 Malaria (2020) Wikipedia (Accessed 2020 Nov 16) https://en.wikipedia.org/wiki/Malaria
3 Barredo, E., & DeGennaro, M. (2020). Not Just from Blood: Mosquito Nutrient Acquisition from Nectar Sources. Trends in parasitology, 36(5), 473–484. https://doi.org/10.1016/j.pt.2020.02.003
4 Nyasembe, V. O., Tchouassi, D. P., Pirk, C., Sole, C. L., & Torto, B. (2018). Host plant forensics and olfactory-based detection in Afro-tropical mosquito disease vectors. PLoS neglected tropical diseases, 12(2), e0006185. https://doi.org/10.1371/journal.pntd.0006185
5 Hien, D. F., Dabiré, K. R., Roche, B., Diabaté, A., Yerbanga, R. S., Cohuet, A., Yameogo, B. K., Gouagna, L. C., Hopkins, R. J., Ouedraogo, G. A., Simard, F., Ouedraogo, J. B., Ignell, R., & Lefevre, T. (2016). Plant- Mediated Effects on Mosquito Capacity to Transmit Human Malaria. PLoS pathogens, 12(8), e1005773. https://doi.org/10.1371/journal.ppat.1005773
6 Ignell, R., & Hill, S. R. (2020). Malaria mosquito chemical ecology. Current opinion in insect science, 40, 6– 10. https://doi.org/10.1016/j.cois.2020.03.008
7 Knols, B. G., & De Jong, R. (1996). Limburger cheese as an attractant for the malaria mosquito Anopheles gambiae s.s. Parasitology today (Personal ed.), 12(4), 159–161. https://doi.org/10.1016/0169-4758(96)10002-8 8 Danquah, I., Bedu-Addo, G., & Mockenhaupt, F. P. (2010). Type 2 diabetes mellitus and increased risk for malaria infection. Emerging infectious diseases, 16(10), 1601–1604. https://doi.org/10.3201/eid1610.100399
9 Fernández-Grandon, G. M., Gezan, S. A., Armour, J. A., Pickett, J. A., & Logan, J. G. (2015). Heritability of attractiveness to

mosquitoes. PloS one, 10(4), e0122716. https://doi.org/10.1371/journal.pone.0122716

10 Ansell, J., Hamilton, K. A., Pinder, M., Walraven, G. E., & Lindsay, S. W. (2002). Short-range attractiveness of pregnant women to Anopheles gambiae mosquitoes. Transactions of the Royal Society of Tropical Medicine and Hygiene, 96(2), 113–116. https://doi.org/10.1016/s0035-9203(02)90271-3

11 Debebe, Y., Hill, S.R., Birgersson, G., Tekie, H., Ignell, R. (2020). Plasmodium falciparum gametocyte- induced volatiles enhance attraction of Anopheles mosquitoes in the field. Malar J 19, 327 (2020). https://doi.org/10.1186/s12936-020-03378-3

12 Robinson, A., Busula, A. O., Voets, M. A., Beshir, K. B., Caulfield, J. C., Powers, S. J., Verhulst, N. O., Winskill, P., Muwanguzi, J., Birkett, M. A., Smallegange, R. C., Masiga, D. K., Mukabana, W. R., Sauerwein, R. W., Sutherland, C. J., Bousema, T., Pickett, J. A., Takken, W., Logan, J. G., & de Boer, J. G. (2018). Plasmodium-associated changes in human odor attract mosquitoes. Proceedings of the National Academy of Sciences of the United States of America, 115(18), E4209–E4218. https://doi.org/10.1073/pnas.1721610115

13 Emami, S. N., Lindberg, B. G., Hua, S., Hill, S. R., Mozuraitis, R., Lehmann, P., Birgersson, G., Borg-Karlson, A. K., Ignell, R., & Faye, I. (2017). A key malaria metabolite modulates vector blood seeking, feeding, and susceptibility to infection. Science (New York, N.Y.), 355(6329), 1076–1080. https://doi.org/10.1126/science.aah4563

14 Lefèvre, T., Gouagna, L. C., Dabiré, K. R., Elguero, E., Fontenille, D., Renaud, F., Costantini, C., & Thomas, F. (2010). Beer consumption increases human attractiveness to malaria mosquitoes. PloS one, 5(3), e9546. https://doi.org/10.1371/journal.pone.0009546

15 Won Jung, J., Baeck, SJ., Perumalsamy, H., Hansson, B.S., Ahn, Y., Wook Kwon, H. (2015) A novel olfactory pathway is essential for fast and efficient blood-feeding in mosquitoes. Sci Rep 5, 13444 (2015). https://doi.org/10.1038/srep13444

16 Wondwosen, B., Birgersson, G., Tekie, H., Torto, B., Ignell, R., Hill, S.R. (2018) Sweet attraction: sugarcane pollen-associated volatiles attract gravid Anopheles arabiensis. Malar J 17, 90 (2018). https://doi.org/10.1186/s12936-018-2245-1

CHAPTER 10

1 Bentz, B.J., Régnière, J., Fettig, C.J., Hansen, E.M., Hayes, J.L., Hicke, J.A., Kelsey, R.G., Negrón, J.F., Seybold, S.J. (2010) Climate Change and Bark Beetles of the Western United States and Canada: Direct and Indirect Effects, BioScience, Volume 60, Issue 8, September 2010, Pages 602–613, https://doi.org/10.1525/bio.2010.60.8.6

2 Santini A, Faccoli M (2015). Dutch elm disease and elm bark beetles: a century of association. iForest 8: 126- 134. - doi: 10.3832/ifor1231-008

3 Holzkurier (translated by Eva Guzely) The dimensions of damage in Europe's forests. timber-online.net. Accessed 2020 Nov 16) Retrieved from https://www.timber-online.net/blog/the-dimensions-of-damage-in- europe-s-forests.html

4 Bark and Wood Boring Beetles of the World (Accessed 2020 Nov 16) www.barkbeetles.org

5 Schmidt, Axel & Zeneli, Gazmend & Hietala, Ari & Fossdal, Carl Gunnar & Krokene, Paal & Christiansen, Erik & Gershenzon, Jonathan. (2005). Induced chemical defences in conifers: Biochemical and molecular approaches to studying their function. Chemical Ecology and Phytochemistry in Forest Ecosystems, 1-28 (2005). 39

6 Wermelinger, B. (2004). Ecology and management of the spruce bark beetle Ips typographus—a review of recent research. Forest Ecology and Management, 202, 67-82

7 Schlyter, Fredrik & Birgersson, Göran & Byers, John & Löfqvist, Jan & Bergström, Gunnar. (1987). Field response of spruce bark beetle, Ips typographus , to aggregation pheromone candidates. Journal of chemical ecology. 13. 701-16. 10.1007/BF01020153

8 Zhang, Qing-He & Song, Li-Wen & Ma, Jian-Hai & Han, Fu-Zhong & Sun, Jianghua. (2009). Aggregation pheromone of a newly described spruce bark beetle, Ips shangrila Cognato and Sun, from China. Chemoecology. 19. 203-210. 10.1007/s00049-009-0026-6

9 Schlyter, F., Birgersson, G., & Leufvén, A. (1989). Inhibition of attraction to aggregation pheromone by verbenone and ipsenol : Density regulation mechanisms in bark beetleIps typographus. Journal of chemical ecology, 15(8), 2263–2277. https://doi.org/10.1007/BF01014114

10 Sauvard D. (2007) General Biology of Bark Beetles. In: Lieutier F., Day K.R., Battisti A., Grégoire JC., Evans H.F. (eds) Bark and Wood Boring Insects in Living Trees in Europe, a Synthesis. Springer, Dordrecht. https://doi.org/10.1007/978-1-4020-2241-8_7

11 Schiebe, C., Hammerbacher, A., Birgersson, G., Witzell, J., Brodelius, P. E., Gershenzon, J., Hansson, B. S., Krokene, P., & Schlyter, F. (2012). Inducibility of chemical defenses in Norway spruce bark is correlated with unsuccessful mass attacks by the spruce bark beetle. Oecologia, 170(1), 183–198. https://doi.org/10.1007/s00442-012-2298-8

12 Netherer, S., Matthews, B., Katzensteiner, K., Blackwell, E., Henschke, P., Hietz, P., Pennerstorfer, J., Rosner, S., Kikuta, S., Schume, H., & Schopf, A. (2015). Do water-limiting conditions predispose Norway spruce to bark beetle attack?. The New phytologist, 205(3), 1128–1141. https://doi.org/10.1111/nph.13166

13 Kandasamy, D., Gershenzon, J., Andersson, M., & Hammerbacher, A. (2019). Volatile organic compounds influence the interaction of the Eurasian spruce bark beetle (Ips typographus) with its fungal symbionts. The ISME Journal, 13, 1788 - 1800. https://doi.org/10.1038/s41396-019-0390-3

14 Anderbrant, O., & Schlyter, F. (1987). Ecology of the Dutch Elm Disease Vectors Scolytus laevis and S. scolytus (Coleoptera: Scolytidae) in Southern Sweden. Journal of Applied Ecology, 24(2), 539-550. doi:10.2307/2403891

15 Schiebe, C., Blaženec, M., Jakuš, R., Unelius, C.R. and Schlyter, F. (2011), Semiochemical diversity diverts bark beetle attacks from Norway spruce edges. Journal of Applied Entomology, 135: 726-737. https://doi.org/10.1111/j.1439-0418.2011.01624.x

16 Weslien, J., & Regnander, J. (2006). The influence of natural enemies on brood production inIps typographus (Col. scolytidae) with special reference to egg-laying and predation by Thanasimus formicarius (Col.: Cleridae). Entomophaga, 37, 333-342. https://doi.org/10.1007/BF02372435

17 Bakke, A., & Kvamme, T. (1981). Kairomone response in Thanasimus predators to pheromone components ofIps typographus. Journal of chemical ecology, 7(2), 305–312. https://doi.org/10.1007/BF00995753

18 Biedermann, P., Müller, J., Grégoire, J. C., Gruppe, A., Hagge, J.,

Hammerbacher, A., Hofstetter, R. W., Kandasamy, D., Kolarik, M., Kostovcik, M., Krokene, P., Sallé, A., Six, D. L., Turrini, T., Vanderpool, D., Wingfield, M. J., & Bässler, C. (2019). Bark Beetle Population Dynamics in the Anthropocene: Challenges and Solutions. Trends in ecology & evolution, 34(10), 914–924. https://doi.org/10.1016/j.tree.2019.06.002

19 Wood, S.L. (1982.) The bark and ambrosia beetles of North and Central America (Coleoptera: Scolytidae), a taxonomic monograph. Great Basin Nat. Mem. 6:1-1356. [304]. https://www.biodiversitylibrary.org/part/248626

CHAPTER 11

1 Drew, Michelle & Harzsch, Steffen & Stensmyr, Marcus & Erland, S. & Hansson, Bill. (2010). A review of the biology and ecology of the Robber Crab, Birgus latro (Linnaeus, 1767) (Anomura: Coenobitidae). Zoologischer Anzeiger - A Journal of Comparative Zoology. 249. 45-67. 10.1016/j.jcz.2010.03.001

2 Christmas Island: A natural wonder https://www.christmas.net.au/

3 Harzsch, S., & Krieger, J. (2018). Crustacean olfactory systems: A comparative review and a crustacean perspective on olfaction in insects. Progress in neurobiology, 161, 23–60. https://doi.org/10.1016/j.pneurobio.2017.11.005

4 Greenaway, P. and Morris, S.(1989). Adaptations to a terrestrial existence by the robber crab, BIRGUS LATRO L.: III. NITROGENOUS EXCRETION. J. Exp. Biol.143, 333-34

5 Drew, M. & Hansson, Bill. (2014). The population structure of Birgus latro (Crustacea: Decapoda: Anomura: Coenobitidae) on Christmas Island with incidental notes on behaviour. The Raffles bulletin of zoology. 150- 161

6 Krieger, J., Grandy, R., Drew, M. M., Erland, S., Stensmyr, M. C., Harzsch, S., & Hansson, B. S. (2012). Giant robber crabs monitored from space: GPS-based telemetric studies on Christmas Island (Indian Ocean). PloS one, 7(11), e49809. https://doi.org/10.1371/journal.pone.0049809

7 Stensmyr, M. C., Erland, S., Hallberg, E., Wallén, R., Greenaway, P., & Hansson, B. S. (2005). Insect-like olfactory adaptations in the

terrestrial giant robber crab. Current biology : CB, 15(2), 116–121. https://doi.org/10.1016/j.cub.2004.12.069

8 Krieger, J., Sandeman, R. E., Sandeman, D. C., Hansson, B. S., & Harzsch, S. (2010). Brain architecture of the largest living land arthropod, the Giant Robber Crab Birgus latro (Crustacea, Anomura, Coenobitidae): evidence for a prominent central olfactory pathway?. Frontiers in zoology, 7, 25. https://doi.org/10.1186/1742-9994-7-25

9 Knaden, M., Bisch-Knaden, S., Linz, J., Reinecke, A., Krieger, J., Erland, S., Harzsch, S., & Hansson, B. S. (2019). Acetoin is a key odor for resource location in the giant robber crab Birgus latro. The Journal of experimental biology, 222(Pt 12), jeb202929. https://doi.org/10.1242/jeb.202929

10 Christmas Island Crab. National Geographic (Accessed 2020 Nov 16) https://www.nationalgeographic.com/animals/invertebrates/c/christmas-island-red-crab/

11 Schildknecht, H., Eßwein, U., Hering, W., Blaschke, C., & Linsenmair, K. (1988). Diskriminierungspheromone der sozialen Wüstenassel Hemilepistus reaumuri / Discriminative Pheromones of the Social Desert Isopod Hemilepistus reaumuri. Zeitschrift für Naturforschung C, 43, 613 - 620

CHAPTER 12

1 Baldwin, I. T., & Schultz, J. C. (1983). Rapid changes in tree leaf chemistry induced by damage: evidence for communication between plants. Science (New York, N.Y.), 221(4607), 277–279. https://doi.org/10.1126/science.221.4607.277

2 Schaller, G. E., & Bleecker, A. B. (1995). Ethylene-binding sites generated in yeast expressing the Arabidopsis ETR1 gene. Science (New York, N.Y.), 270(5243), 1809–1811. https://doi.org/10.1126/science.270.5243.1809

3 Pare, P. W., & Tumlinson, J. H. (1999). Plant volatiles as a defense against insect herbivores. Plant physiology, 121(2), 325–332

4 Huang, W., Gfeller, V., & Erb, M. (2019). Root volatiles in plant-plant interactions II: Root volatiles alter root chemistry and plant-herbivore interactions of neighbouring plants. Plant,

cell & environment, 42(6), 1964–1973. https://doi.org/10.1111/pce.13534

5 Nagashima, A., Higaki, T., Koeduka, T., Ishigami, K., Hosokawa, S., Watanabe, H., Matsui, K., Hasezawa, S., & Touhara, K. (2018). Transcriptional regulators involved in responses to volatile organic compounds in plants. The Journal of Biological Chemistry, 294, 2256 - 2266. doi:10.1074/jbc.RA118.005843

6 The networked beauty of forests - Suzanne Simard. TED-Ed (Accessed 2020 Nov 16) https://ed.ted.com/lessons/the-networked-beauty-of-forests-suzanne-simard

7 Markovic, D., Colzi, I., Taiti, C., Ray, S., Scalone, R., Gregory Ali, J., Mancuso, S., & Ninkovic, V. (2019). Airborne signals synchronize the defenses of neighboring plants in response to touch. Journal of experimental botany, 70(2), 691–700. https://doi.org/10.1093/jxb/ery375

8 Heil, M., & Silva Bueno, J. C. (2007). Within-plant signaling by volatiles leads to induction and priming of an indirect plant defense in nature. Proceedings of the National Academy of Sciences of the United States of America, 104(13), 5467–5472. https://doi.org/10.1073/pnas.0610266104

9 Christensen, S. A., Nemchenko, A., Borrego, E., Murray, I., Sobhy, I. S., Bosak, L., DeBlasio, S., Erb, M., Robert, C. A., Vaughn, K. A., Herrfurth, C., Tumlinson, J., Feussner, I., Jackson, D., Turlings, T. C., Engelberth, J., Nansen, C., Meeley, R., & Kolomiets, M. V. (2013). The maize lipoxygenase, ZmLOX10, mediates green leaf volatile, jasmonate and herbivore-induced plant volatile production for defense against insect attack. The Plant journal : for cell and molecular biology, 74(1), 59–73. https://doi.org/10.1111/tpj.12101

10 Clavijo McCormick, A., Irmisch, S., Reinecke, A., Boeckler, G. A., Veit, D., Reichelt, M., Hansson, B. S., Gershenzon, J., Köllner, T. G., & Unsicker, S. B. (2014). Herbivore-induced volatile emission in black poplar: regulation and role in attracting herbivore enemies. Plant, cell & environment, 37(8), 1909–1923. https://doi.org/10.1111/pce.12287

11 Sukegawa, S., Shiojiri, K., Higami, T., Suzuki, S., & Arimura, G. I. (2018). Pest management using mint volatiles to elicit resistance in

soy: mechanism and application potential. The Plant journal : for cell and molecular biology, 96(5), 910–920. https://doi.org/10.1111/tpj.14077

12 Coll-Aráoz, M.V., Hill, J.G., Luft-Albarracin, E., Virla, E.G., & Fernandez, P.C. (2020). Modern Maize Hybrids Have Lost Volatile Bottom-Up and Top-Down Control of Dalbulus maidis, a Specialist Herbivore. Journal of Chemical Ecology, 1 - 10. https://doi.org/10.1007/s10886-020-01204-3

13 Oluwafemi, S., Dewhirst, S. Y., Veyrat, N., Powers, S., Bruce, T. J., Caulfield, J. C., Pickett, J. A., & Birkett, M. A. (2013). Priming of Production in Maize of Volatile Organic Defence Compounds by the Natural Plant Activator cis-Jasmone. PloS one, 8(6), e62299. https://doi.org/10.1371/journal.pone.0062299

14 Rasmann, S., Köllner, T. G., Degenhardt, J., Hiltpold, I., Toepfer, S., Kuhlmann, U., Gershenzon, J., & Turlings, T. C. (2005). Recruitment of entomopathogenic nematodes by insect-damaged maize roots. Nature, 434(7034), 732–737. https://doi.org/10.1038/nature03451

15 Köllner, T. G., Held, M., Lenk, C., Hiltpold, I., Turlings, T. C., Gershenzon, J., & Degenhardt, J. (2008). A maize (E)-beta-caryophyllene synthase implicated in indirect defense responses against herbivores is not expressed in most American maize varieties. The Plant cell, 20(2), 482–494. https://doi.org/10.1105/tpc.107.051672

CHAPTER 13

1 Sprengel CK. 1793. Das entdeckte Geheimniss der Natur im Bau und in der Befruchtung der Blumen. Berlin, Germany: Friedrich Vieweg

2 Darwin, C. (1862).On the various contrivances by which British and foreign orchids are fertilised by insects. First edition. London

3 Darwin, C. (1877).On the various contrivances by which orchids are fertilised by insects. Second edition, revised. London

4 Jersakova, Jana & Johnson, Steven & Kindlmann, Pavel. (2006). Mechanisms and evolution of deceptive pollination in orchids. Biological reviews of the Cambridge Philosophical Society. 81. 219-35. 10.1017/S1464793105006986

5 Schiestl, Florian & Ayasse, Manfred & Paulus, Hannes & Löfstedt, Christer & Hansson, Bill & Ibarra, Fernando & Francke, Wittko. (2000). Sex pheromone mimicry in the early spider orchid (Ophrys sphegodes): patterns of hydrocarbons as the key mechanism for pollination by sexual deception. Journal of comparative physiology. A, Sensory, neural, and behavioral physiology. 186. 567-74. 10.1007/s003590000112

6 Schiestl, Florian & Ayasse, Manfred & Paulus, Hannes & Löfstedt, Christer & Hansson, Bill & Ibarra, Fernando & Francke, Wittko. (1999). Orchid pollination by sexual swindle [5]. Nature. 399. 421-421. 10.1038/20829

7 Ayasse, M., Schiestl, F. P., Paulus, H. F., Löfstedt, C., Hansson, B., Ibarra, F., & Francke, W. (2000). Evolution of reproductive strategies in the sexually deceptive orchid Ophrys sphegodes: how does flower-specific variation of odor signals influence reproductive success?. Evolution; international journal of organic evolution, 54(6), 1995–2006. https://doi.org/10.1111/j.0014-3820.2000.tb01243.x

8 Stensmyr, M. C., Urru, I., Collu, I., Celander, M., Hansson, B. S., & Angioy, A. M. (2002). Pollination: Rotting smell of dead-horse arum florets. Nature, 420(6916), 625–626. https://doi.org/10.1038/420625a

9 Angioy, A. M., Stensmyr, M. C., Urru, I., Puliafito, M., Collu, I., & Hansson, B. S. (2004). Function of the heater: the dead horse arum revisited. Proceedings. Biological sciences, 271 Suppl 3(Suppl 3), S13–S15. https://doi.org/10.1098/rsbl.2003.0111

10 Stökl, J., Strutz, A., Dafni, A., Svatos, A., Doubsky, J., Knaden, M., Sachse, S., Hansson, B. S., & Stensmyr, M. C. (2010). A deceptive pollination system targeting drosophilids through olfactory mimicry of yeast. Current biology : CB, 20(20), 1846–1852. https://doi.org/10.1016/j.cub.2010.09.033

11 Stökl, Johannes & Brodmann, Jennifer & Dafni, Amots & Ayasse, Manfred & Hansson, Bill. (2010). Smells like aphids: orchid flowers mimic aphid alarm pheromones to attract hoverflies for pollination. Proceedings. Biological sciences / The Royal Society. 278. 1216-22. 10.1098/rspb.2010.1770

12 Gemeno, C., Yeargan, K.V. & Haynes, K.F. (2000) Aggressive Chemical Mimicry by the Bolas Spider Mastophora hutchinsoni:

Identification and Quantification of a Major Prey's Sex Pheromone Components in the Spider's Volatile Emissions. J Chem Ecol 26, 1235–1243 (2000). https://doi.org/10.1023/A:1005488128468

13 Keesey, I. W., Koerte, S., Khallaf, M. A., Retzke, T., Guillou, A., Grosse-Wilde, E., Buchon, N., Knaden, M., & Hansson, B. S. (2017). Pathogenic bacteria enhance dispersal through alteration of Drosophila social communication. Nature communications, 8(1), 265. https://doi.org/10.1038/s41467-017-00334-9

CHAPTER 14

1 Trivedi, D. K., Sinclair, E., Xu, Y., Sarkar, D., Walton-Doyle, C., Liscio, C., Banks, P., Milne, J., Silverdale, M., Kunath, T., Goodacre, R., & Barran, P. (2019). Discovery of Volatile Biomarkers of Parkinson's Disease from Sebum. ACS central science, 5(4), 599–606. https://doi.org/10.1021/acscentsci.8b00879

2 HeroRAT Magawa - PDSA Gold Medal PDSA.org.uk Retrieved from https://www.pdsa.org.uk/what-we- do/animal-awards-programme/pdsa-gold-medal/magawa

3 Bromenshenk, J. J., Henderson, C. B., Seccomb, R. A., Welch, P. M., Debnam, S. E., & Firth, D. R. (2015). Bees as Biosensors: Chemosensory Ability, Honey Bee Monitoring Systems, and Emergent Sensor Technologies Derived from the Pollinator Syndrome. Biosensors, 5(4), 678–711. https://doi.org/10.3390/bios5040678

4 Manjunatha, D H & Chua, Lee Suan. (2017). Advancement of sensitive sniffer bee technology. TrAC Trends in Analytical Chemistry. 97. 10.1016/j.trac.2017.09.006

5 Wilson, Alphus. (2012). Review of Electronic-Nose Technologies and Algorithms to Detect Hazardous Chemicals in the Environment. Procedia - Technology. 1. 453-463. 10.1016/j.protcy.2012.02.101

6 Gardner, J.W., Bartlett, P.N. (1994) A brief history of electronic noses, Sensors and Actuators B: Chemical, Volume 18, Issues 1–3, 1994, Pages 210-211, ISSN 0925-4005, https://doi.org/10.1016/0925-4005(94)87085-3

7 Hu, Wenwen & Wan, Liangtian & Jian, Yingying & Ren, Cong & Jin, Ke & Su, Xinghua & Bai, Xiaoxia & Haick, Hossam &

Yao, Mingshui & Wu, Weiwei. (2018). Electronic Noses: From Advanced Materials to Sensors Aided with Data Processing. Advanced Materials Technologies. 10.1002/admt.201800488

8 Arshak, K., Moore, E.G., Lyons, G.R., Harris, J., & Clifford, S. (2004). A review of gas sensors employed in electronic nose applications. Sensor Review, 24, 181-198. DOI 10.1108/02602280410525977

9 Snow, R. W., Rowan, K. M., Lindsay, S. W., & Greenwood, B. M. (1988). A trial of bed nets (mosquito nets) as a malaria control strategy in a rural area of The Gambia, West Africa. Transactions of the Royal Society of Tropical Medicine and Hygiene, 82(2), 212–215. https://doi.org/10.1016/0035-9203(88)90414-2

10 Knols, B., Farenhorst, M., Andriessen, R., Snetselaar, J., Suer, R., Osinga, A., Knols, J. Deschietere, J., Lyimo, I., Kessy, S., Mayagaya, V., Sperling, S., Cordel, M., Sternberg, E., Hartmann, P., Mnyone, L., Rose, A., Thomas, M. (2016). Eave tubes for malaria control in Africa: An introduction. Malaria Journal. 15.10.1186/s12936-016-1452-x

11 Dawit, Mengistu & Hill, Sharon & Birgersson, Göran & Tekie, Habte & Ignell, Rickard. (2020). Malaria mosquitoes acquire and allocate cattle urine to enhance life history traits. 10.1101/2020.08.24.264309

12 Raty, L., Drumont, A., De Windt, N., Grégoire, J (1995) Mass trapping of the spruce bark beetle Ips typographus L.: traps or trap trees?, Forest Ecology and Management, Volume 78, Issues 1–3, 1995, Pages 191- 205, ISSN 0378-1127, https://doi.org/10.1016/0378-1127(95)03582-1

13 Khan, Zeyaur & Midega, Charles & Pittchar, Jimmy & Pickett, John & Bruce, Toby. (2011). Push-pull technology: a conservation agriculture approach for integrated management of insect pests, weeds and soil health in Africa UK government's Foresight Food and Farming Futures project. International Journal of Agricultural Sustainability. 9. 162-170. 10.3763/ijas.2010.0558

14 Saini, R. K., Orindi, B. O., Mbahin, N., Andoke, J. A., Muasa, P. N., Mbuvi, D. M., Muya, C. M., Pickett, J. A., & Borgemeister, C. W. (2017). Protecting cows in small holder farms in East Africa from tsetse flies by mimicking the odor profile of a non-host bovid. PLoS neglected tropical diseases, 11(10), e0005977. https://doi.org/10.1371/journal.pntd.0005977

15 Daum, R. F., Sekinger, B., Kobal, G., & Lang, C. J. (2000). Riechprüfung mit "sniffin' sticks" zur klinischen Diagnostik des Morbus Parkinson [Olfactory testing with "sniffin' sticks" for clinical diagnosis of Parkinson disease]. Der Nervenarzt, 71(8), 643–650. https://doi.org/10.1007/s001150050640

Index

M

N

O

V

W

Y

Z

Born in Sweden, the neuroethologist Bill S. Hansson served as Vice President of the Max Planck Society and is currently Director of the Max Planck Institute for Chemical Ecology in Jena, Germany, and an honorary professor at Friedrich Schiller University. His research centres on the question of how plants and insects communicate through scent.